Mathematical Methods for Life Sciences

Mathematical Methods for Life Sciences introduces calculus, and other key mathematical methods, to students from applied sciences (biology, biotechnology, chemistry, pharmacology, material science, etc.). Special attention is paid to real-world applications, and for every concept, many concrete examples are provided. This book does not aim to enable students to prove theorems and construct elaborate proofs, but rather to leave students with a clear understanding of the practical mathematics behind the power of optimization, dynamical systems and all the predictive tools these theories give rise to.

Features

- No prerequisites beyond high school algebra and geometry

- Could serve as the primary text for a first-year course in mathematical methods for biology, biotechnology or other life sciences

- Easy to read: the students may skip all the proofs and go directly to key examples and applications

Cinzia Bisi is a professor of Geometry at the Department of Mathematics and Computer Sciences at the University of Ferrara, Italy. She has wide experience in teaching mathematics and statistics to students in the Department of Life Sciences. She has an interest in the areas of pure and applied mathematics.

Rita Fioresi is a professor of Geometry at the FaBiT Department at the University of Bologna, Italy. She has written textbooks in linear algebra, and her research interests are primarily in the areas of pure and applied mathematics.

Mathematical Methods for Life Sciences

Cinzia Bisi
University of Ferrara, Italy

Rita Fioresi
University of Bologna, Italy

CRC Press

Taylor & Francis Group

Boca Raton London New York

CRC Press is an imprint of the
Taylor & Francis Group, an **informa** business

A CHAPMAN & HALL BOOK

First edition published 2024
by CRC Press
2385 NW Executive Center Drive, Suite 320, Boca Raton FL 33431

and by CRC Press
4 Park Square, Milton Park, Abingdon, Oxon, OX14 4RN

CRC Press is an imprint of Taylor & Francis Group, LLC

ISBN: 9781032362298 (hbk)
ISBN: 9781032380582 (pbk)
ISBN: 9781003343288 (ebk)

DOI: 10.1201/9781003343288

Typeset in Minion
by codeMantra

This Work is devoted to Carla, Ruggero,
Roma e Renzo.

Contents

Acknowledgments

We thank, first of all, our students, who, in the course of the years, have helped us to understand how to better present the theory and applications of differential calculus.

We thank Prof. Ermanno Lanconelli for his precious advice on the order of exposition and Prof. Francesco Faglioni for his technical help.

We also thank the many teaching assistants who, over the years, assisted our students, proposing exercises and helping them toward the solution: Filippo Caleca, Eugenia Celada, Gaia Fontana, Michela Lapenna, Martina Pepiciello, Luigi Scurto, Claudio Severi, Simone Tentori, Simone Tibaldi and Pierpaolo Vecchi. We also remember Giulio Ribani.

We finally thank the FaBiT Department (Unibo) and the Mathematics Department (Unife) for their hospitality and technical support while writing this textbook.

Introduction

This textbook comes from our experience in teaching calculus for biology, biotechnology and chemistry students.

Our aim is to give solid foundations for differential and integral calculus and, at the same time, to show how powerful this beautiful theory is by providing applications to biology, chemistry and, in general, applied sciences.

In all chapters, together with a rigorous treatment including the key definitions, statements of results and their proofs, we always look at concrete examples in which the power of the theory is manifest, for example, the study of the time evolution of an epidemic or the variation of the concentration of a drug in the bloodstream.

We suggest the following reading order for our textbook. A first introductory course can start from Chapter 1 or, alternatively, directly from Chapter 2 and end at Chapter 4 (integrals). Chapter 7 with elementary statistics and probability is independent from the rest of our treatment. Chapters 5 and 6 discuss differential equations, and hence, they can be skipped if time is pressing.

We summarize the guide to the reading of our textbook with the following diagram.

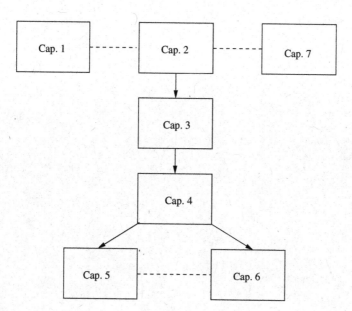

CHAPTER 1

Functions in Applied Sciences

1.1 THE CONCEPT OF FUNCTION

We start with the most important definition of this chapter.

Definition 1.1.1 A function f between two sets A and B is a law assigning to each element a of the set A, called the *domain* of f, a single element b of the set B, called the *codomain* of f. The element b is called the *image* of a and it is denoted by $f(a)$. The set of all images $f(a)$ for each a in A is called the *image* of f, and it is denoted by $\text{Im}(f)$.

We are particularly interested in functions, with domain and codomain contained in the set of real numbers $\mathbb{R}$.

Example 1.1.2 Let us consider the law that associates with every positive real number r, the area of the circle of radius r with $r > 0$. We see that it is a function, because we can describe this law as:

$$f : \mathbb{R}^+ \longrightarrow \mathbb{R}, \quad f(r) = \pi r^2$$

The domain of f is given by the positive real numbers $\mathbb{R}^+$, while the codomain consists of all real numbers. However, we note that the image of f is given by positive numbers only: $\text{Im}(f) = \mathbb{R}^+$. Hence, equivalently we can write $f : \mathbb{R}^+ \longrightarrow \mathbb{R}^+$.

DOI: 10.1201/9781003343288-1

We now come to the concept of a graph of a function. We always assume f to be defined on a domain D consisting of real numbers: $f(x) \in \mathbb{R}$ for each x in D.

Definition 1.1.3 Let $f : D \longrightarrow \mathbb{R}$ be a function defined on a domain $D \subset \mathbb{R}$. The *graph* of f, denoted by $G(f)$, is the set of ordered pairs $(x, f(x))$ with x in D. In formulas, we write:

$$G(f) := \{(x, y) \mid x \in D, \, y = f(x)\}$$

We represent $G(f)$ in the Cartesian plane, by placing the value x on the x-axis and the value $f(x)$ on the y-axis.

Example 1.1.4 Given the function $f(x) = -2x^2$, we want to determine its domain and its graph in the Cartesian plane. The domain is the set of real numbers $\mathbb{R}$, which, from now on, we identify with the x-axis of the Cartesian plane. We notice that the value of $f(x)$ is always negative or zero. Hence, we have that the image of f is contained in the set of real numbers less or equal to zero. It is not difficult to check if it coincides with such a set. Indeed, for every $y \leq 0$, we have that $x = \sqrt{-(1/2)y}$ is such that $f(x) = -2(\sqrt{-y/2})^2 = y$.

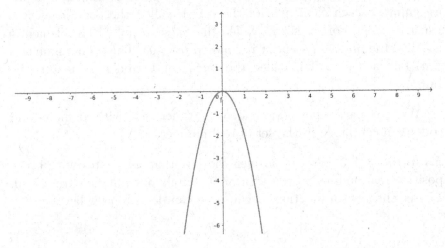

1.2 LINEAR FUNCTIONS

We now examine the *linear functions*, which are of fundamental importance, since we can effectively approximate most physical or biological systems using linear functions.

Definition 1.2.1 A function $f : D \longrightarrow \mathbb{R}$ is *linear*, if $f(x) = mx + c$ for suitable coefficients $m \neq 0$, c in $\mathbb{R}$.

We immediately see that both the domain and the image of a linear function coincide with the real numbers.

The graph of a linear function is a line in the Cartesian plane and it is described by the equation $y = mx + c$. The number m is called the *slope* and, as we shall see, is of great importance.

We have that:

- If $m = 0$, the line is horizontal.

- The term c gives information about the points where the line intersects the Cartesian axes. In fact, if we substitute $y = 0$ in the equation of the line $y = mx + c$, we get the point of intersection of the line with the x-axis. Similarly, if we set $x = 0$, we get the point of intersection with the y-axis. These points are $(-\frac{c}{m}, 0)$ and $(0, c)$, respectively.

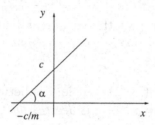

Let us now see the geometric meaning of the slope m. Let us consider a linear function $f(x) = mx + c$ and the line described by its graph $y = mx + c$.

Consider the two points on the line: $P = (x_1, y_1)$, with $f(x_1) = y_1$ and $Q = (x_2, y_2)$, with $f(x_2) = y_2$. So, we have:

$$f(x_1) = mx_1 + c, \qquad f(x_2) = mx_2 + c$$

We obtain the following expression for the slope m:

$$f(x_2) - f(x_1) = y_2 - y_1 = m(x_2 - x_1), \qquad \rightarrow \qquad m = \frac{y_2 - y_1}{x_2 - x_2}$$

in terms of the coordinates of P and Q.

If α denotes the angle between the line and the x-axis, this formula gives:

$$m = \tan(\alpha)$$

(see also Section 1.7).

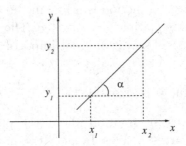

It is important to notice that the computation of m is *independent* from the choice of the points P and Q.

We now want to collect some important information on the linear function $f(x) = mx + c$, by looking at the slope of the line $y = mx + c$.

- If $m > 0$, we have that the angle α is between 0 and $\frac{\pi}{2}$; hence, the function $f(x) = mx + c$ is *increasing*, that is, $f(x_1) < f(x_2)$, if $x_1 < x_2$.

- If $m < 0$, we have that the angle α is between $-\frac{\pi}{2}$ and 0; hence, the function $f(x) = mx + c$ is *decreasing*, that is, $f(x_1) > f(x_2)$, if $x_1 < x_2$.

- If $m = 0$, we have $\alpha = 0$, so the line is *horizontal*. In this case, the function f is constant.

Let us see how to compute the line passing through two given points.

Example 1.2.2 We want to determine the equation of the line passing through the points $A = (1, 2)$ and $B = (2, 5)$ of the Cartesian plane. We first consider the equation $y = mx + c$, and then, we write the conditions for the points A and B to belong to the line. In other words, we substitute the coordinates of the points A and B in the equation $y = mx + c$ to find m and c. We obtain:

$$2 = m + c, \quad 5 = 2m + c$$

Solving the linear system, we have $m = 3$ and $c = -1$. We can alternatively compute m and c directly, using the formula for m we derived above:

$$m = \frac{y_2 - y_1}{x_2 - x_1} = 3, \qquad c = 2 - 3 \cdot 1 = -1$$

Notice that we used the A point to obtain c, after we computed m, but we may have alternatively used B.

Let us see a more concrete example of application of linear functions.

Example 1.2.3 Centigrade and Farenheit scales. Let C express the temperature in centigrades (or Celsius) and F in Farenheit. We see experimentally that we can express centigrade units in terms of Farenheit units using a linear function. We want to determine such function, by knowing that $0°C = 32°F$ and each centigrade degree (unit) is 9/5 the Farenheit degree. Hence, an interval of temperature in centigrade degrees is 5/9, the same interval expressed in Farenheit degrees.

We set up the linear function:

$$C = mF + c$$

The equation $0°C = 32°F$ immediately gives us $0 = 32m + c$, that is, $c = -32m$. Hence, $C = m(F - 32)$. This information gives us:

$$C = (5/9)(F - 32).$$

Linear functions also appear in mixing problems.

Example 1.2.4 We want to prepare a mousse containing pear puree and plain yogurt. Pear puree contains 0.3% of protein, while plain yogurt contains 11% of protein. We want to calculate in what percentage we should mix pear puree and yogurt so that the mousse contains 1% of protein.

We set the following linear equation:

$$\frac{(0.3)x}{100} + \frac{(11)y}{100} = \frac{1(x+y)}{100}$$

Clearing common denominators, we get:

$$(0,3)x + 11y = x + y$$

From which

$$(0,7)x = 10y$$

that is to say

$$y = \frac{(0,7)}{10}x = \frac{7}{100}x.$$

Hence, the amount of yogurt should be 7% of the pear puree one.

1.3 POLYNOMIAL FUNCTIONS

We now take into exam polynomial functions, which are of great importance in physics and biology.

Definition 1.3.1 We call $f : D \longrightarrow \mathbb{R}$ a *polynomial function*, if we can express it through a polynomial, that is, $f(x) = a_n x^n + \cdots + a_1 x + a_0$ for suitable real coefficients $a_n, \ldots, a_1, a_0$.

We immediately see that the domain D is given by all real numbers, while the image, depending on the degree of the polynomial, may or may not coincide with the real numbers, as we shall see later on.

The graph of a polynomial function represents a curve in the plane. We focus now our attention on the case of degree 2, that is:

$$f(x) = ax^2 + bx + c, \quad a \neq 0$$

The graph of a degree 2 polynomial function, is a *parabola*:

$$y = ax^2 + bx + c$$

Let us summarize some simple geometric observations.

- If $a > 0$, then the parabola has positive concavity, while if $a < 0$, the concavity is negative.

- The vertex of the parabola is $V = (-\frac{b}{2a}, -\frac{b^2 - 4ac}{4a})$.

- If $a > 0$, the y coordinate of the vertex represents the minimum of the function $f(x) = ax^2 + bx + c$, while, if $a < 0$, it represents the maximum.

- The point $(0, c)$ is the intersection of the parabola with the y-axis.

- The solutions of the equation $ax^2 + bx + c = 0$ give the intersections of the parabola with the x-axis.

More precisely, let $\Delta = b^2 - 4ac$ be the discriminant of the equation $ax^2 + bx + c = 0$ and

$$x_1 = \frac{-b + \sqrt{b^2 - 4ac}}{2a}, \qquad x_2 = \frac{-b - \sqrt{b^2 - 4ac}}{2a}$$

the solutions of the equation.

- If $\Delta > 0$, the parabola intersects the x-axis in the points with x coordinate x_1 or x_2. If the parabola has positive concavity, the graph is above the x-axis for $x < x_1$ and $x > x_2$.

- If $\Delta = 0$, the parabola intersects the x-axis at one point only. If the concavity is positive, the graph is above the x-axis. If the concavity is negative, the graph is below the x-axis.

- If $\Delta < 0$, the parabola does not intersect the x-axis. If the concavity is positive, the graph is above the x-axis, and if negative, it is below the x-axis.

In summary:

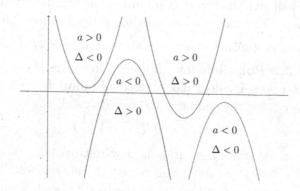

Let us now look at some applications of great physical and biological interest.

Example 1.3.2 The law of falling bodies:

The law describing the falling of a body, subject to gravity, expresses the position $s(t)$ of the body in free fall from an initial position s_0 and subject at an initial velocity v_0:

$$s(t) = s_0 + v_0 t - (1/2)gt^2$$

as function of the time t. The minus sign in $-(1/2)gt^2$ is due to the acceleration of gravity $g = 9.8$ m/s^2, giving the force of gravity, which we assume to be "downward". Clearly, this is a convention, but in this text we will always adopt it.

For example, if we throw a ball upward at a speed of 6 m/s, we have $s_0 = 0$ and $v_0 = 6$. Hence, the height, or position, of the ball is given by:

$$s(t) = 6t - (9.8/2)t^2 = 6t - 4.9t^2$$

This law is expressed by a polynomial function of degree 2. Notice that the graph of this function is a parabola, but it does not represent the trajectory of the object, but only the variation of its height $s(t)$ as time varies. In other words, the ball moves vertically to reach a maximum height and then returns to the ground.

We know that the maximum of a parabola with negative concavity is obtained at its vertex, which has coordinates

$$(t_{max}, s_{max}) = (6/9.8, 36/19.6) = (0.61, 1.84).$$

Hence, the maximum height obtained by the ball is the y coordinate of the vertex, that is, 1.84 m. If we want to compute after how many seconds the ball reaches the maximum height, we have to look at the first coordinate. The ball reaches maximum height after 0.61 seconds.

Let us look at another example, taken from biology: Poiselle's law.

Example 1.3.3 Poiselle's law:
Poiselle's law gives the change in the speed of blood flow, inside a blood vessel of radius R, as a function of r, the distance from the center of the blood vessel:

$$v(r) = k(R^2 - r^2)$$

where k is a constant. This parabolic law tells us that the maximum of the blood velocity occurs at the vertex of the parabola, that is, for $r = 0$, which corresponds to center of the blood vessel. The minimum, on the contrary, is reached on the walls of the blood vessel, that is, for $r = R$.

1.4 RATIONAL FUNCTIONS AND ALGEBRAIC FUNCTIONS

In this section, we define important types of functions, obtained from polynomial functions performing some simple operations.

Definition 1.4.1 We say that $f : D \longrightarrow \mathbb{R}$ is a *rational function*, if $f(x)$ is the quotient of two polynomials, that is:

$$f(x) = \frac{a_n x^n + \cdots + a_0}{b_m x^m + \cdots + b_0}$$

The domain D consists of the values of x for which the denominator is non-zero, i.e., the values of x in $\mathbb{R}$ for which $b_m x^m + \cdots + b_0 \neq 0$.

Let us see an example of domain of these functions.

Example 1.4.2 Consider the function:

$$f(x) = \frac{3x - 2x^3}{x^4 - 1}$$

We want to determine the domain of f. We need to compute the values of x for which the denominator is zero:

$$x^4 - 1 = (x^2 - 1)(x^2 + 1) = 0 \implies x = \pm 1$$

Hence, the domain is given by the set of real numbers except the values ± 1. We write $D = \mathbb{R} \setminus \{\pm 1\}$, or equivalently:

$$D = \{x \in \mathbb{R} \mid x \neq \pm 1\}$$

Definition 1.4.3 We say that $f : D \longrightarrow \mathbb{R}$ is an *algebraic function*, if $f(x)$ is obtained through a finite number of algebraic operations from a polynomial function. By algebraic operations, we mean taking quotient and raising to the n-th power, where n can be a rational number, i.e., $n = p/q$ (p and q integers).

The following are algebraic functions:

$$\frac{1}{x - 2}, \qquad \sqrt{x^2 + 3}, \qquad \frac{(x - 1)^{1/3}}{x^3 - 2}, \qquad (x^4 - 3x - 5)^{5/6}$$

The following are *not* algebraic functions:

$$\frac{e^x}{x - 2}, \qquad \sqrt{x^2 + \sin(x)}, \qquad \frac{(x - 1)^\pi}{x^3}$$

To compute the domain, it is necessary to take into account the denominator, but also to see when the radicand is positive, in case of even n, for nth roots (like square root, fourth root, etc.).

Example 1.4.4 We want to compute the domain D for the following algebraic functions.

1. $\frac{1}{x-2}$, $D = \mathbb{R} \setminus \{2\}$.

2. $\sqrt{x^2 + 3}$, $D = \mathbb{R}$, (notice: $x^2 + 3$ is always positive).

3. $\frac{(x-1)^{1/3}}{x^3-2}$, $D = \mathbb{R} \setminus \{2^{1/3}\}$.

   In fact, we can decompose the denominator as

   $$x^3 - 2 = (x - 2^{1/3})(x^2 + 2^{1/3}\, x + 2^{2/3})$$

   which is zero for $x = 2^{1/3}$. Notice that the numerator is a cubic root, and hence, there are no extra conditions.

4. $(x^2 - 3x - 5)^{5/6}$,

   $$D = \{x \in \mathbb{R} \mid x \le (3 - \sqrt{29})/2,\ x \ge (3 + \sqrt{29})/2\}$$

   Since the root is even, we must impose the positivity condition for the argument.

1.5 THE EXPONENTIAL AND LOGARITHMIC FUNCTIONS

The exponential function is of fundamental importance in applied sciences, as we will see in the description of the Malthusian model in the next section.

Definition 1.5.1 We define the exponential function, for $a \in \mathbb{R}^+$, as $f : \mathbb{R} \longrightarrow \mathbb{R}$, $f(x) = a^x$.

The domain of the exponential function is given by $\mathbb{R}$, while the graph is the following:

- for $a > 1$:

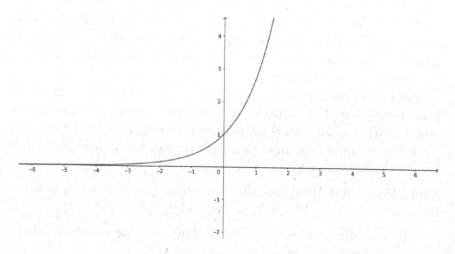

• for $a < .1$:

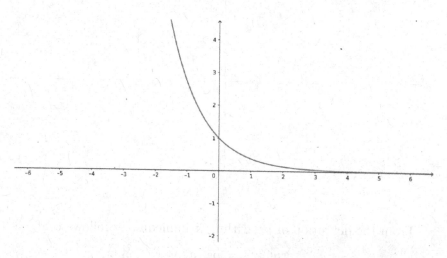

The domain must take into account the case in which the exponent of a is also a function. Let us see an example.

Example 1.5.2 We want to determine the domain of the function

$$f(x) = 2^{\frac{1}{x-1}}$$

As for the exponential function, we have that the domain is given by the real numbers; however, since the exponent is a function defined for $x \neq 1$, we have that the domain of f is $\mathbb{R} \setminus \{1\}$.

We are particularly interested in the exponential function when $a = e$. The number e is the *Napier's constant*, also called *Euler's number*. This is a number between 2 and 3, whose importance will be clarified in later chapters.

Let us now examine the logarithmic function. Recall that, given two positive real numbers a and b, the logarithm with respect to the base a of b, written as $\log_a(b)$, is the exponent to which a must be raised to get b. For example, the logarithm with respect to the base 2 of 8 is 3: $\log_2 8 = 3$, since $2^3 = 8$.

Definition 1.5.3 We define the logarithmic function with respect to the base $a \in \mathbb{R}^+$, $f : D \longrightarrow \mathbb{R}$, as $f(x) = \log_a(x)$.

The domain D of the logarithmic function consists of all strictly positive real numbers, and the graph is given by:

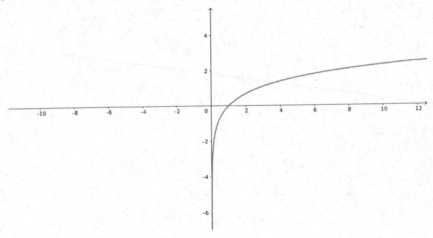

From the definition of logarithm, it immediately follows that:

$$a^{\log_a x} = \log_a a^x = x$$

Hence, we have that the logarithmic function is the *inverse function* of the exponential function: that is, if we apply the exponential function to x and then the logarithmic function (or vice versa), we get x, provided we are within their respective domains.

As with the exponential, we are particularly interested in the logarithm with respect to the base e, but the formula of change of base allows us to write any logarithm with respect to the base e and vice versa:

$$\log_e x = \frac{\log_a x}{\log_a e}, \qquad \log_a x = \frac{\log_e x}{\log_e a}$$

For clarity, we do not write the base a, when $a = e$.

1.6 MALTHUSIAN LAW

Let us now see a concrete application of the exponential function: the *Malthusian law*. Suppose we have a population of $N(t)$ individuals, increasing at a fixed rate R.

For example, we want to consider a population of patients affected by COVID and we assume that every day every patient infects R ones.

In Malthusian law, R is constant, while very often in biological phenomena, R itself depends on time,[1] thus making the mathematical description of the phenomena more involved. We will consider only the case of constant R. We have:

Days	0	1	2	3
Inf. Ind.	N_0	$N_0 R$	$N_0 R^2$	$N_0 R^3$

The function expressing the number of individuals of a given population at time t is an exponential function:

$$N(t) = N_0 R^t, \qquad \text{Malthusian law}$$

In our example, $N(t)$ is the number of individuals affected by COVID at time t, expressed in days. Assume to have $R = 1.2$ and initially 0.5 million of infected individuals. After a week, we have:

$$N(7) = 0.5 \cdot (1.2)^7 \cong 1.79$$

that is, about 1.8 million COVID-infected individuals. This is actually based on real data coming from Italian official numbers in winter 2021, when the COVID omicron variant was spreading. It is important to remember that this is an approximate model, giving only a rough estimate of the size of the infected population over a short period of time. For example, it does not take into consideration neither the individuals who recover nor the deceased ones, contributing to a

---

[1] This is the R_t factor in COVID epidemics.

negative change in the population. We will see, later on, more accurate models that can take into account such important factors.

Let us see, in the next example, how to apply Malthusian equation to other biology questions.

Example 1.6.1 Suppose we have a bacterial population, which duplicates itself every 12 hours. If we initially have $N_0 = 10^2$ bacteria, we want to determine the number of bacteria after two days and the function $N(t)$, which describes the time evolution of the bacterial population.

Let us see the total number of bacteria expressed in a table:

Hours	0	12	24	36	48
No. bacteria	N_0	$2\,N_0$	$4\,N_0$	$8\,N_0$	$16\,N_0$

The number of bacteria after 2 days is $16N_0 = 16 \cdot 10^2$. The law describing such number over time is:

$$N(t) = N_0 2^t$$

This is an exponential function, with base $R = 2$. The variable t expresses the time in the chosen time unit, in this case an interval of 12 hours, called *cycle*. We notice that, if we express time in hours t_{hours}, we can get t in cycle units, by dividing by 12: $t = t_{hours}/12$.

We must write all time variables with respect to the same unit of time.

The preferred base for the exponential function in mathematics is $a = e$, where e is a *transcendental number*, whose value is between 2 and 3. The word "transcendental" means that e is neither an irrational nor a rational number. Furthermore, we cannot obtain it by solving an equation with rational coefficients, as we would for $\sqrt{2}$, which is the solution of $x^2 - 2 = 0$. The reason why this base is better than any other one will be clear later, when we study derivatives.

We recall the base change formula for the exponential function:

$$a^x = e^{x \log(a)}, \qquad e^x = a^{x \log_a(e)}$$

Hence, we can solve any problem of population growth according to Malthusian laws with the base e. Let us see an example.

Example 1.6.2 Suppose we have 30g of Uranium 238 decaying into Thorium 233 with half-life of $t_{1/2} = 4.5 \times 10^5$ years. This means that, after a time interval of $t_{1/2}$, we have half of the initial quantity of Uranium 238. We can easily see that the half-life does not depend on the initial quantity of radioactive material.

We can immediately set up the Malthusian equation:

$$N(t) = 30(1/2)^{t/(4.5 \times 10^5)} \tag{1.1}$$

where the time t is expressed in years. To obtain the cycles, we must divide by 4.5×10^5.

However, we can also equivalently express this equation as:

$$N(t) = 30e^{-\lambda t} \tag{1.2}$$

We can get the parameter λ via the two equations (1.1) and (1.2).

$$30(1/2)^{t/4.5 \times 10^5} = 30e^{-\lambda t} \quad \Longrightarrow \quad -\frac{t}{t_{1/2}}\log 2 = -\lambda t$$

from which $\lambda = \log(2)/t_{1/2}$. Hence:

$$\lambda = 0.7/(4.5 \cdot 10^5) = 1.54 \times 10^{-6}$$

We can now immediately determine, for example, how many grams of Uranium 238 we will have after 10,000 years:

$$N(10000) = 30e^{-\lambda t} = 30e^{-1.54 \cdot 10^{-6} \cdot 10^4} = 30e^{0.0154} = 29.54$$

In the previous example, we have observed that the half-life $t_{1/2}$ of a population following a Malthusian law, does not depend on the initial number of individuals. Let us see this fact more explicitly. We have:

$$N(t_{1/2}) = N_0 e^{-\lambda t_{1/2}} = N_0/2 \quad \Longrightarrow \quad t_{1/2} = \frac{\log 2}{\lambda}$$

where λ is constant.

1.7 ELEMENTARY TRIGONOMETRIC FUNCTIONS

In this section, we want to briefly recall a few facts regarding elementary trigonometric functions: sine, cosine and tangent. Let α be an angle in a right triangle. By definition, cosine and sine of α are given, respectively, by the quotient of the side adjacent to α and the hypotenuse and the quotient of the opposite side and the hypotenuse, respectively, as we see in the figure below.

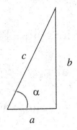

$$\cos(\alpha) = \frac{a}{c}, \qquad \sin(\alpha) = \frac{b}{c}$$

It is easy to prove, using Thales theorem, that the definitions of sine and cosine are independent of the chosen right triangle. For this reason, we represent cosine and sine of an angle using a right triangle with hypotenuse of unit length. In this way, we have that the cosine of an angle is given by the length of the adjacent side to α, while the sine of an angle is given by the length of the opposite side.

To draw and better visualize cosine and sine, we use the unit circle: it is a circle with a radius of length 1 and centered at the origin of the Cartesian plane.

As we see from the picture, we can immediately determine the cosine and sine of an angle α: they are the coordinates (x, y) of the corresponding point on the circle (Figure 1.1).

By Pythagoras Theorem, we see immediately the *Pythagorean trigonometric identity*

$$\cos^2 \alpha + \sin^2 \alpha = 1$$

We conventionally set α positive, when we move counterclockwise and negative, when we move clockwise. The units of measure for angles are degrees and radians, where in radians we have that 2π corresponds to 360°. We see in the table below the most common angles expressed

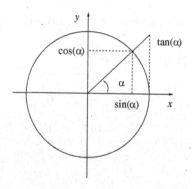

Figure 1.1 Unit Circle

both in degrees and in radians. Recall that in trigonometry and more generally in mathematics, the expression of an angle in radians is the most used.

Degrees	Radians		Degrees	Radians
30	$\frac{\pi}{6}$		210	$\frac{7}{6}\pi$
45	$\frac{\pi}{4}$		240	$\frac{4}{3}\pi$
60	$\frac{\pi}{3}$		270	$\frac{3}{2}\pi$
90	$\frac{\pi}{2}$			
120	$\frac{2}{3}\pi$		300	$\frac{5}{3}\pi$
150	$\frac{5}{6}\pi$		330	$\frac{11}{6}\pi$
180	π		360	2π

The unit circle is divided into four sectors or quadrants:

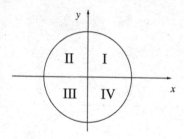

We see explicitly the value of cosine and sine for the angles in each quadrant.

First and second quadrants

α	cos	sin
0	1	0
$\frac{\pi}{6}$	$\frac{\sqrt{3}}{2}$	$\frac{1}{2}$
$\frac{\pi}{4}$	$\frac{\sqrt{2}}{2}$	$\frac{\sqrt{2}}{2}$
$\frac{\pi}{3}$	$\frac{1}{2}$	$\frac{\sqrt{3}}{2}$

α	cos	sin
$\frac{\pi}{2}$	0	1
$\frac{2}{3}\pi$	$-\frac{1}{2}$	$\frac{\sqrt{3}}{2}$
$\frac{3}{4}\pi$	$-\frac{\sqrt{2}}{2}$	$\frac{\sqrt{2}}{2}$
$\frac{5}{6}\pi$	$-\frac{\sqrt{3}}{2}$	$\frac{1}{2}$

Third and fourth quadrants

α	cos	sin
π	-1	0
$\frac{7}{6}\pi$	$-\frac{\sqrt{3}}{2}$	$-\frac{1}{2}$
$\frac{5}{4}\pi$	$-\frac{\sqrt{2}}{2}$	$-\frac{\sqrt{2}}{2}$
$\frac{4}{3}\pi$	$-\frac{1}{2}$	$-\frac{\sqrt{3}}{2}$

α	cos	sin
$\frac{3}{2}\pi$	0	-1
$\frac{5}{3}\pi$	$\frac{1}{2}$	$-\frac{\sqrt{3}}{2}$
$\frac{7}{4}\pi$	$\frac{\sqrt{2}}{2}$	$-\frac{\sqrt{2}}{2}$
$\frac{11}{6}\pi$	$\frac{\sqrt{3}}{2}$	$-\frac{1}{2}$

Let us summarize all information:

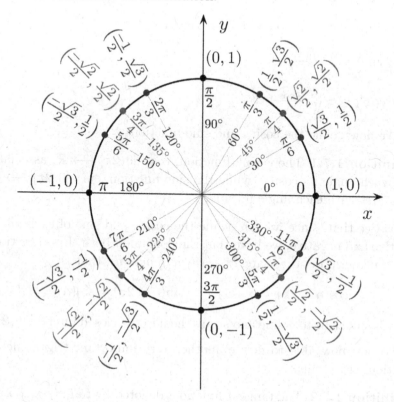

We now write some useful formulas, which we do not prove here, that come directly from the definitions. We invite the student to think about the cosine and sine of an angle α and the other angles obtained from it by simple operations (as $-\alpha$). By geometric observations on the unit circle, it is easy to see all of the given formulas. Similar reasoning shows that we need to know only the values of the angles expressed in first table, namely, the angles of the first quadrant, as the values pertaining to the other quadrants are obtained from them.

- $\cos(-\alpha) = \cos\alpha$

- $\sin(-\alpha) = -\sin\alpha$

- $\cos(\alpha + \pi/2) = -\sin(\alpha)$

- $\sin(\alpha + \pi/2) = \cos(\alpha)$

- $\cos(\alpha + \beta) = \cos \alpha \cos \beta - \sin \alpha \sin \beta$

- $\sin(\alpha + \beta) = \sin \alpha \cos \beta + \sin \beta \cos \alpha$

- $\sin \frac{\alpha}{2} = \pm \sqrt{\frac{1 - \cos \alpha}{2}}$

- $\cos \frac{\alpha}{2} = \pm \sqrt{\frac{1 + \cos \alpha}{2}}$

We now define the cosine and sine functions.

Definition 1.7.1 The cosine function, $\cos : [0, 2\pi] \longrightarrow \mathbb{R}$, associates with each angle its cosine, while the sine function, $\sin : [0, 2\pi] \longrightarrow \mathbb{R}$, associates with each angle its sine.

Notice that, since we can define the cosine and sine of each angle, positive and negative, we have that the domain of both these functions is $\mathbb{R}$. Furthermore, these are 2π *periodic functions*, that is:

$$\cos(\alpha + 2\pi) = \cos(\alpha), \qquad \sin(\alpha + 2\pi) = \sin(\alpha)$$

So, it is sufficient to define and study them in the closed interval $[0, 2\pi]$.

We are now able to define another very important trigonometric function: the tangent.

Definition 1.7.2 The tangent function, denoted by $\tan : (-\pi/2, \pi/2) \longrightarrow \mathbb{R}$, associates with an angle the quotient of its sine and cosine: $\tan(x) = \sin(x)/\cos(x)$.

The name "tangent" comes from the fact that we can visualize the value of the tangent of an angle by looking at the line tangent to the unit circle at the point $(1, 0)$. The domain of the tangent function is given by $\mathbb{R} \setminus \{\pi/2 + k\pi\}$, that is, we remove the zeros of the denominator $\cos(x)$. Since tan is periodic, we are interested in its behavior only for x between $-\pi/2$ and $\pi/2$.

Another important function in trigonometry is the *cotangent function*, which should not be confused with the inverse of the tangent function, which we will see later.

Definition 1.7.3 The cotangent function, denoted by $\cot : (0, \pi) \longrightarrow \mathbb{R}$, associates with an angle the quotient of its cosine and sine of the angle: $\cot(x) = \frac{\cos(x)}{\sin(x)} = \frac{1}{\tan(x)}$.

We now want to draw the graphs of the trigonometric functions that we have defined. We invite the student to convince himself/herself, by examining the unit circle, that the graphs we draw are correct. For sine and cosine, we have:

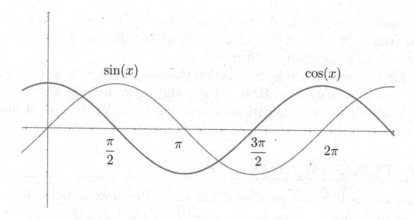

while the graph of the tangent is:

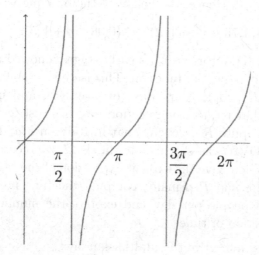

We also consider the inverse functions of the elementary trigonometric functions cos, sin, tan called, respectively, *arcsine*, *arccosine* and *arctangent* and denoted by arccos, arcsin and arctan.

Recall that the inverse function of a function f is a function that applied to f gives identity, naturally, taking x in the domain of f.

Definition 1.7.4 The arccosine function, denoted by arccos : $[-1, 1] \longrightarrow [0, \pi]$, associates with the number x the (unique) angle in the interval $[0, \pi]$ whose cosine is x.

So, if we apply the function arccos to the quantity $\cos(\alpha)$, we get the angle α.

Similarly, we define the functions arcsin : $[-1, 1] \longrightarrow [-\pi/2, \pi/2]$ and arctan : $\mathbb{R} \longrightarrow [-\pi/2, \pi/2]$ as the functions that applied to $\sin(\alpha)$ and $\tan(\alpha)$, respectively, return α.

Equivalently, arcsin is the function that applied to x gives the angle whose sine is equal to x. Hence, if $x = \sin(\alpha)$, we have $\arcsin(x) = \alpha$ and similarly for the other trigonometric functions. We can also define, in a similar way, the arctangent function arctan.

1.8 EXERCISES WITH SOLUTIONS

1.8.1 A brick falls from a building and, after 0.98 seconds, hits a passerby 1.75m tall. Determine the height of the building.

Solution. The law of falling bodies tells us that: $s(t) = s_0 + v_0 t - (1/2)gt^2$ with $s_0 = 0$, $v_0 = 0$. Hence, the height of the building is:

$$|s(0.98)| + 1.75\text{m} = +(1/2)9.8 \cdot (0.98)^2 \text{m} + 1.75\text{m} = 6.46\text{m}$$

1.8.2 The COVID outbreak in its early stages followed a Malthusian law, that is, an exponential function. This is due to the fact that every person infects R people every day (on average), and initially there are no factors that reduce the infection rate like restrictive measures, deaths, etc. Compute R, knowing that in Italy on March 5, 2020, we had 769 new COVID cases, while on March 12, we recorded 2653 new COVID cases. Then, give the explicit expression of the Malthusian law, both with base e and R. Finally, compute the time needed to reach 20,000 infected people per day and express the number of infected people as a function of time.

Solution. Let us first write the Malthusian law:

$$N(t) = 769 \cdot R^7 = 769 e^{7 \log(R)}$$

We impose that, in a seven-day period, the number of daily COVID cases increases to 2653:

$$2653 = 769 e^{t \log(R)} \qquad \Longrightarrow \qquad e^{7 \log(R)} = 3.45$$

that is:

$$\log(R) = \log(3.45)/7 = 0.177, \quad \Longrightarrow \quad R = 1.2$$

To determine the time t_*, corresponding to $20,000$ daily infections, we write:

$$N(t_*) = 769e^{t_* \log(R)} = 769e^{0.177t_*} = 20,000$$

We immediately obtain:

$$0.177t_* = \log(20,000/769) = 3.24 \quad \Longrightarrow \quad t_* = 3.24/0.177 \cong 18.4$$

That is, we reach 20,000 daily infections in about 18 days.

1.8.3 We sprinkle oranges with a certain pesticide that covers their peel. However, after 4 days, we see a reduction by 50% of the amount of toxic substance in the orange peel. By law, oranges cannot be sold until we measure less than 10% of the initial toxic substance. We further assume that its quantity in the oranges follows a Malthusian law. How many days do we have to wait to be able to sell the oranges?

Solution. The percentage of pesticide in the orange peel follows the Malthusian law:

$$N(t) = N(0)e^{-\lambda t}$$

where $N(t)$ is the quantity of pesticide and the half-life is $t_{1/2} = 4$ days. This means that:

$$N(0)e^{-\lambda t_{1/2}} = N(0)/2$$

So, we have:

$$-\lambda t_{1/2} = \log(1/2) \quad \Longrightarrow \quad \lambda = \frac{\log(2)}{t_{1/2}} = 0.173$$

To find out after how many days oranges are safe to eat, we must impose that the amount of pesticide in the oranges is 10% of the initial amount:

$$N(t_{\text{comm}}) = N(0)e^{-\lambda t_{\text{comm}}} = N(0)/10$$

We have:

$$t_{\text{comm}} = -\log(1/10)/0.173 \cong 13.3 \,\text{giorni}$$

So, we need to wait at least 13.3 to safely consume and sell the oranges.

We now set the problem in an equivalent way, describing the Malthusian law through the formulation $N(t) = N(0)R^t$. We have:

$$N(t) = N(0)(1/2)^{t_c} = N(0)(1/2)^{t/4}$$

where t_c is expressed in cycles and t in days. We now impose that the quantity of pesticide is reduced to $1/10$ of the initial amount:

$$N(0)(1/2)^{t_{comm}/4} = N(0)/10$$

We obtain:

$$(1/2)^{t_{comm}/4} = 1/10 \quad \Longrightarrow \quad (t_{comm}/4)\log(1/2) = \log(1/10)$$

Hence:

$$t_{comm} = \frac{4\log(1/10)}{\log(1/2)} \cong 13.3 \, \text{giorni}$$

We, thus, reobtain the previous result.

1.8.4 In a right triangle, a side measures 2 cm and forms an angle of 30° with the hypotenuse. Find the hypotenuse and the length of other side.

Solution. Recall the definition of cosine and sine: The cosine of an angle is the quotient of the adjacent side and the hypotenuse. Hence, we can immediately compute the hypotenuse of the given triangle:

$$\frac{2}{\cos(30^o)} = \frac{4}{\sqrt{3}}$$

Since the sine of an angle is the quotient between the opposite side and the hypotenuse, we have that:

$$\sin(30^o) = \frac{\text{opposite side}}{\frac{4}{\sqrt{3}}}$$

Hence, since $\sin(30^o) = 1/2$, the length of the side is given by: $\frac{2}{\sqrt{3}}$.

1.9 SUGGESTED EXERCISES

1.9.1 For each of the following laws, determine whether or not it is a function, and in case it is, find the domain and the codomain.

1. The law that associates each person with his/her siblings.

2. The law that associates each person with his/her mother.

3. The law that associates each positive integer with its divisors.

4. The law that associates each positive integer with its cube.

1.9.2 For each pair of points A and B, compute the equation of the line through A and B and, if possible, its slope.

1. $A = (1, 0)$, $B = (1, -1)$.

2. $A = (1, 2)$, $B = (3, -1)$.

3. $A = (-1, 4)$, $B = (1/2, -1)$.

4. $A = (1/3, 0)$, $B = (1, -2)$.

1.9.3 Write the line through the point $(2, 3)$ and with slope $m = 3$.

1.9.4 For each of the following functions, determine the domain.

1.
$$f(x) = \frac{2x^2 + x + 2}{x^2 - 1}$$

2.
$$f(x) = \frac{e^{1/x}}{x^2 + 1}$$

3.
$$f(x) = \sin(1/x)\cos(\sqrt{x})$$

4.
$$f(x) = \log(x)e^{2x-7}$$

5.
$$f(x) = \log(1 + x + x^2)$$

6.
$$f(x) = \sqrt{1 - 2x + x^2}$$

7.
$$f(x) = e^{\frac{x+1}{x^2+3x-2}}$$

8.
$$f(x) = \frac{1}{5 + e^{2x}}$$

9.
$$f(x) = \frac{e^{-x}}{\tan(x)}$$

10.
$$f(x) = \frac{x - 2}{(x^2 - 3)^{1/3}}$$

11.
$$f(x) = \log\left(\frac{x^2 - 36}{x + 4}\right)$$

12.
$$f(x) = \log|x + 3|$$

13.
$$f(x) = e^{\frac{x-3}{x+2}}$$

14.
$$f(x) = \frac{x - 1}{\sqrt{x^2 + x - 2}}$$

15.
$$f(x) = \log\left(\frac{x - 1}{x^2 - 9}\right)$$

16.
$$f(x) = 3x - \sqrt{x^2 - 4}$$

17.
$$f(x) = e^{\sqrt{x^2 - 4}}$$

18.
$$f(x) = e^{|x + 2|}$$

1.9.5 Compute the following logarithms *without* the use of a calculator:

1. $\log_3(27)$

2. $\log_5(1/625)$

3. $\log_{2/3}(\sqrt{3/2})$ •

4. $\log_{\sqrt{7}} 49$

1.9.6 Determine the base a of the following logarithms:

1. $\log_a(81) = 4$

2. $\log_a(1/16) = 4$

3. $\log_a 1/\sqrt{2} = 1/2$

4. $\log_a 49 = -4$

1.9.7 The dosage $D(n)$ of a pediatric antihistaminic medicine follows the law:

$$D(n) = \frac{n+2}{24}A$$

where n is the age of the child in years ($n \geq 2$) and A is the dosage for adults.

If for a two-year-old child we have an 80 mg dosage, what will the dosage be for a four-year-old child?

1.9.8 Suppose we have two bottles of chlorine disinfectant, one with a 2% solution of active principle and the other with a 10% solution. In what proportions do we have to mix the two solutions to obtain a 4% solution?

1.9.9 A ball is dropped from a height of 450 m. After how many seconds does it reach the ground? [Note: $g = 9.8$ m/s^2].

1.9.10 We drop a ball from a building 15 m tall, imparting an initial downward velocity of 5 m/s.

a. What will its speed be just before it hits the ground?

b. How long will it take to reach the ground?

    c. Answer to questions (a) and (b) if the ball is thrown upward imparting an initial velocity of 5 m/s.

    d. Plot the height of the ball as a function of time.

1.9.11 On a construction site, the only cable that supports a freight elevator breaks on top of a 100 m high building.

    a. How fast does the elevator hit the ground?

    b. How long does it fall?

    c. What is its speed at mid-height?

    d. When it reaches the mid-height, how long has it been falling?

1.9.12 Let A and B be the alleles of the cystic fibrosis gene and suppose their percentage among the population is p and $1 - p$ ($0 \leq p \leq 1$), respectively. Individuals carry both alleles (heterozygotes) with the percentage $2p(1 - p)$. What is the value of p for which we have the higher percentage of heterozygotes?

1.9.13 The half-life of ibuprofen in the blood is about 2 hours. If a patient takes 400 mg of ibuprofen, compute its amount in the blood after 45 minutes, 90 minutes and 24 hours.

1.9.14 If the amount of ozone in the atmosphere decreases by 15% per year, how long does it take to get to half of the starting amount?

1.9.15 To estimate the age of some archaeological artifact, we use a method based on the percentage of Carbon-14, with respect to the total carbon in the artifact. Carbon-14 is a radioactive isotope of carbon, which has a half-life of 5730 years. If an artifact has 27 % of Carbon-14, what is its age?

1.9.16 A bacterial colony follows a Malthusian law and quadruples every 5 days. Write the number $N(t)$ of bacteria as time varies, knowing that we initially have about 4000 bacteria. Find out how many bacteria we have after 23 days.

1.9.17 $N(t) = 4.5(1.33)^t$ gives the (Malthusian) model of population growth in a South American nation from 1815 onward, where t is expressed in decades and $N(t)$ is given in units of 10,000 people.

1. Determine whether the population duplicated itself between 1815 and 1840.

2. Determine if there is a 20-year period, in which the population has doubled.

1.9.18 With a chemical process, it is possible to duplicate a certain substance at each cycle. If after five cycles we have about 0.00118 g of substance, how much did we initially have? Determine how many cycles are needed to have 1g of substance starting from 0.00118 g of substance.

1.9.19 Starting with certain amount of DNA, through the polymerase process we can be duplicate it at each cycle. If after 20 cycles we have about 0.00112 g, what is the amount of DNA that we initially had?

1.9.20 Aspirin is eliminated by the kidneys at a rate of 50 % per hour.

a. Find how many hours are needed to have only 10% of the quantity of aspirin initially administered.

b. Can the table represent the amount of aspirin in the body at a given time?

Quantity	Time (in min)
500 mg	0
419 mg	6
307 mg	12
230 mg	18

1.9.21 To eliminate 50% of a certain antibiotic from the blood, the body takes about 4 hours. Does this amount of time depend on the milligrams of antibiotic in the blood? Motivate clearly your answer.

1.9.22 The growth chart of golden retriever puppies is given by:

Weight in kg	months
5.3	2
9.1	3
12.7	4
15.8	5

1. Determine if it is (approximately) a linear, quadratic or exponential function.

2. Give a puppy's weight estimate at 40 days based on the law found in point 1.

1.9.23 The speed of water in a river is measured as a function of depth and reported by the following table:

Depth in meters	Speed
2.0	1.11
2.6	1.42
3.3	1.39
4.6	1.39
5.9	1.14
7.3	0.91

1. Determine if it follows a linear, quadratic, exponential law.

2. Give an estimate of the speed at the depth of 4 m based on the law of point 1.

1.9.24 Suppose we have the following table, derived from experimental data:

Concentration (micromolar) of the substrate S	Reaction rate V micromolar/h
0.12	0.22
0.54	0.72
1.24	1.2
1.8	1.4
2.7	1.6

Find a line $y = mx + c$ that approximates the data with $y = 1/S$ e $x = 1/V$.

1.9.25 In a right triangle, the hypotenuse measures 3 cm and forms an angle of 45° with one of the sides. Find the lengths of both sides.

1.9.26 Compute the area and perimeter of a right triangle knowing that one of the sides is 4 cm long and forms an angle of 60° with the hypotenuse.

1.9.27 Determine the tangent of the angle whose cosine measures $1/5$.

1.9.28 Determine the tangent of α, knowing that $\sin(\alpha) = 0.2$.

1.9.29 Find the sine of the angle with tangent equal to 3.

1.9.30 Determine the cosine of the angle $\alpha + \pi/2$, knowing that $\cos(\alpha) = 0.6$, without the use of a calculator.

1.9.31 Solve the following trigonometric equations and inequalities:

1. $\sin x + \cos x = 1$

2. $\sin x > \frac{\sqrt{3}}{2}$

3. $\sin^2 x > \frac{1}{2}$

4. $\cos^2 x < \frac{1}{2}$

1.9.32 Determine $\arccos(1/2)$, $\arcsin(0)$, $\arctan(1)$ without the help of a calculator.

1.9.33 Determine x if $\arccos(x) = \arcsin(x)$. Is x unique?

CHAPTER 2

Limits and Derivatives

2.1 LIMITS

The concept of limit represents the foundation of differential and integral calculus: it tells us how the value $f(x)$ of a function f changes as the variable x approaches a value x_0.

For example, consider the linear function $f(x) = x + 3$. We see that $f(x)$ approximates the number 5, as x approaches the value $x_0 = 2$. We can easily see it, if we substitute the value 2 for x: $f(2) = 2 + 3 = 5$ and we look at the graph of the line $y = x + 3$. Indeed, for values of x close to the point $x_0 = 2$ on the x-axis, the corresponding values of the function f are very close to $f(2) = 5$.

x	$f(x)$
2	5
2.1	5.1
1.9	4.9

DOI: 10.1201/9781003343288-2

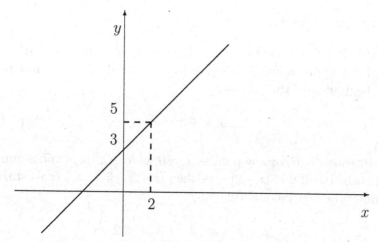

As we shall see, we need the definition of limit to understand the values of the function f at x "near" x_0, when we cannot evaluate the function f at x_0.

Definition 2.1.1 Let $f : D \longrightarrow \mathbb{R}$ and suppose that f is defined at all points of an open interval containing x_0, but not necessarily in x_0. We say that the real number L is the *limit* of $f(x)$ for x approaching x_0, if, for every $\epsilon > 0$, there exists $\delta > 0$ such that, if $0 < |x - x_0| < \delta$, we have:

$$|f(x) - L| < \epsilon$$

We write in formulas:

$$\lim_{x \to x_0} f(x) = L$$

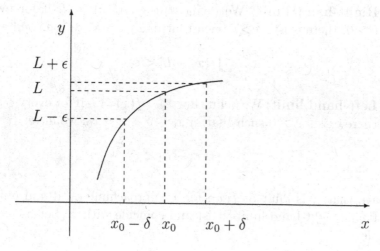

Let us see an example.

Example 2.1.2 We want to check that the limit of the function $f(x) = x + 3$, for x approaching 2, is equal to 5, as we intuitively anticipated at the beginning of this section:

$$\lim_{x \to 2} x + 3 = 5$$

We must show that, for any number $\epsilon > 0$, we can always find a number $\delta > 0$, such that if $0 < |x - 2| < \delta$, then $|x + 3 - 5| < \epsilon$. Let us start by examining the last inequality:

$$|x - 2| < \epsilon \iff 2 - \epsilon < x < 2 + \epsilon$$

So if $\delta = \epsilon$, we have $0 < |x - 2| < \delta$.

We invite the reader to check by exercise, using the reasoning of the previous example, that:

$$\lim_{x \to x_0} x = x_0, \qquad \lim_{x \to x_0} k = k \qquad (2.1)$$

where $k \in \mathbb{R}$ is a constant.

We can also define the right-hand limit and the left-hand limit of a function at x_0.

- **Right-hand limit:** We define $\lim_{x \to x_0^+} f(x) = L$ if for every $\epsilon > 0$, there exists $\delta > 0$ such that, if $x_0 < x < x_0 + \delta$, we have:

$$|f(x) - L| < \epsilon$$

- **Left-hand limit:** We define $\lim_{x \to x_0^-} f(x) = L$ if for every $\epsilon > 0$, there exists $\delta > 0$ such that, if $x_0 - \delta < x < x_0$, we have:

$$|f(x) - L| < \epsilon$$

We have that L is limit of $f(x)$, for x approaching x_0, if and only if the left and right-hand limits exist and coincide with L. Let us see an example.

Example 2.1.3 Let us consider the *step function* $f : \mathbb{R} \longrightarrow \mathbb{R}$:

$$f(x) = \begin{cases} 0 & x < 0 \\ 1 & x \geq 0 \end{cases}$$

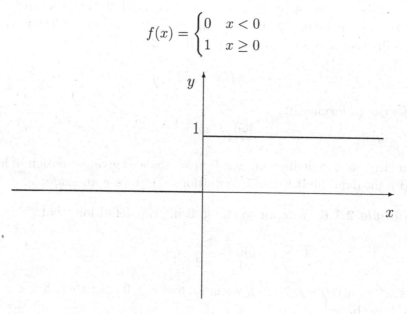

We leave to the reader the easy check that:

$$\lim_{x \to 0^-} f(x) = 0 \neq 1 = \lim_{x \to 0^+} f(x)$$

Hence, the limit for x approaching 0 of f does not exist.

It is really important to notice that, when calculating the limit of a function f at a point, we are not interested in the value of the function at the point. In fact, in the definition of limit for x approaching x_0, f *may not be defined at the point* x_0.

We now state an important result: the uniqueness of the limit. For its proof, see the Appendix 2.13.

Theorem 2.1.4 Uniqueness of limit: *Let* $f : D \longrightarrow \mathbb{R}$ *and assume f to be defined at all points x in an open interval containing x_0, but not necessarily at x_0. If the limit of f for x approaching x_0 exists, then it is unique.*

We conclude this section with other important definitions. We invite the reader to do the exercises at the end of the chapter for many other cases.

Definition 2.1.5 Let $f : D \longrightarrow \mathbb{R}$ and assume f to be defined at all the points x in an open interval containing x_0, but not necessarily at x_0. We say the limit of $f(x)$ for x approaching x_0 is infinite if, for every $M > 0$, there exists $\delta > 0$ such that, if $0 < |x - x_0| < \delta$, we have:

$$f(x) > M$$

We write in formulas:

$$\lim_{x \to x_0} f(x) = +\infty$$

Similarly, we can define $\lim_{x \to x_0} f(x) = -\infty$ and give the notion of left and right-hand limit for such definitions. Let us see an example.

Example 2.1.6 We want to check, using the definition, that:

$$\lim_{x \to 1^+} \frac{1}{x - 1} = +\infty$$

Taking an arbitrary $M > 0$, we must find $\delta > 0$ such that, if $1 < x < 1 + \delta$, we have:

$$\frac{1}{x - 1} > M$$

Hence:

$$\frac{1}{x - 1} > M \iff x - 1 < 1/M \iff x < 1 + 1/M$$

So, if we take $\delta = 1/M$, we have that, for $1 < x < 1 + 1/M$, we get $1/(x - 1) > M$.

Now let us see the definition of limit, when x goes to infinity.

Definition 2.1.7 Let $f : D \longrightarrow \mathbb{R}$ and assume f to be defined for all the points $x > N$, where N is a real number. We say that L is the limit of $f(x)$ for x approaching infinity if, for every $\epsilon > 0$, there exists $M > 0$ $(M > N)$, such that, if $x > M$, we have:

$$|f(x) - L| < \epsilon$$

We write in formulas:

$$\lim_{x \to +\infty} f(x) = L$$

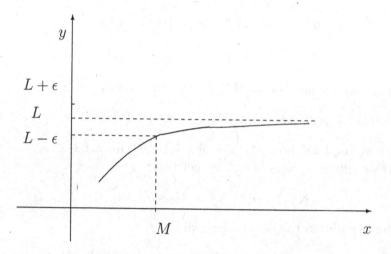

2.2 PROPERTIES OF LIMITS AND STANDARD LIMITS

In this section we outline some important properties of limits (see Appendix 2.13 for more details).

Proposition 2.2.1 Properties of limits: *Consider two functions* $f, g : D \longrightarrow \mathbb{R}$ *and suppose* $\lim_{x \to x_0} f(x)$, $\lim_{x \to x_0} g(x)$ *exist and are finite. Then:*

1. $\lim_{x \to x_0} (f(x) + g(x)) = \lim_{x \to x_0} f(x) + \lim_{x \to x_0} g(x)$.

2. $\lim_{x \to x_0} (f(x) \cdot g(x)) = \lim_{x \to x_0} f(x) \cdot \lim_{x \to x_0} g(x)$.

3. $\lim_{x \to x_0} (f(x)/g(x)) = \lim_{x \to x_0} f(x) / \lim_{x \to x_0} g(x)$, *where* $\lim_{x \to x_0} g(x) \neq 0$.

4. $\lim_{x \to x_0} (f(x))^{p/q} = (\lim_{x \to x_0} f(x))^{p/q}$, $p, q \in \mathbb{Z}$, $q \neq 0$.

Proof. We prove property (1), the other properties can be proven in a similar way and we leave them as an exercise.

We define the numbers L and M as:

$$\lim_{x \to x_0} f(x) = L, \qquad \lim_{x \to x_0} g(x) = M$$

We need to find, for every $\epsilon > 0$, a number $\delta > 0$ such that, if $0 < |x - x_0| < \delta$, we have $|f(x) + g(x) - L - M| < \epsilon$. By hypothesis, for each $\epsilon_1, \epsilon_2 > 0$, we have $\delta_1, \delta_2 > 0$, such that:

$$\text{if} \quad 0 < |x - x_0| < \delta_1, \quad |f(x) - L| < \epsilon_1,$$

$$\text{if} \quad 0 < |x - x_0| < \delta_2, \quad |g(x) - M| < \epsilon_2.$$

Taking δ as the smaller number between δ_1 and δ_2, we have:

$$|f(x) + g(x) - L - M| \leq |f(x) - L| + |g(x) - M| < \epsilon_1 + \epsilon_2$$

where we used the property $|a + b| \leq |a| + |b|$, for $a, b \in \mathbb{R}$.

If we choose $\epsilon_1 = \epsilon_2 = \epsilon/2$, we get that:

$$|f(x) + g(x) - L - M| < \epsilon_1 + \epsilon_2 = \epsilon$$

so that we obtain the desired inequality. $\qquad\qquad\qquad \square$

The first two statements of this proposition apply also in the more general case in which the limits are both infinite with the same sign, for example, $+\infty$. We leave to the reader the simple modification of this proof to take into account this case. Later on, we will see in detail how to compute the limit of quotients of functions, whose limits are infinite.

The previous proposition allows us to compute immediately some simple limits.

Example 2.2.2   1. Compute $\lim_{x \to 3} 2x^2 - 7$. As we know from (2.1), $\lim_{x \to 3} x = 3$ and $\lim_{x \to 3} 2 = 2$. Using the properties of limits, we have that:

$$\lim_{x \to 3} 2x^2 - 7 = 2 \cdot 3^2 - 7 = 11$$

2. Compute the $\lim_{x \to -1} \sqrt{3x^3 + 8}$. Using the properties of limits, we have immediately:

$$\lim_{x \to -1} \sqrt{3x^3 + 8} = \sqrt{3(-1)^3 + 8} = \sqrt{5}$$

Proposition 2.2.1 does not always help to compute limits, because sometimes it leads to *indeterminate forms*, that is, to expressions like

$$\frac{0}{0}, \qquad \frac{\infty}{\infty}, \qquad 0 \cdot \infty, \qquad +\infty - \infty$$

Let us see how to solve them in some simple cases.

Example 2.2.3 We want to compute the limit:

$$\lim_{x \to 2} \frac{x^2 - 4}{x - 2} = \frac{0}{0}$$

However, we note that the numerator is a polynomial, which can be factored: $x^2 - 4 = (x - 2)(x + 2)$. So, we write:

$$\lim_{x \to 2} \frac{x^2 - 4}{x - 2} = \lim_{x \to 2} x + 2 = 4$$

We cannot always solve indeterminate forms in such an easy way. We now give a proposition for some limits of particular interest that lead to the indeterminate form $\frac{0}{0}$.

Proposition 2.2.4 Standard limits: *We have:*

1. $\lim_{t \to 0} \frac{\sin(t)}{t} = 1$.

2. $\lim_{t \to 0} \frac{e^x - 1}{t} = 1$.

3. $\lim_{t \to 0} \frac{\log(x + 1)}{t} = 1$.

Proof. Let us see the proof of (1). Let us consider an angle t on the unit circle.

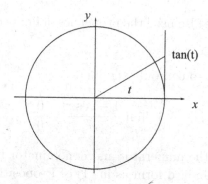

We see that:

$$\sin(t) < t < \tan(t)$$

Now we observe that we can divide all terms in the inequality by $\sin t$, so that we get:

$$1 < \frac{t}{\sin(t)} < \frac{1}{\cos(t)}$$

Taking the inverse of all (positive) terms, we have:

$$1 > \frac{\sin(t)}{t} > \cos(t)$$

By the Comparison Theorem in the Appendix 2.13, we get the result.

The proof of (2) and (3) are considerably more complicated and, in particular, (2) can also be used as the definition of the number e. We send the interested reader to more advanced textbooks as [3]. □

Let us see how we can use the previous proposition to compute some limits.

Example 2.2.5   1. Let us consider the limit:

$$\lim_{x \to 0} \frac{\sin 2x}{e^x - 1} = \frac{0}{0}$$

In order to compute it, we want to make use of the standard forms. We have:

$$\lim_{x \to 0} \frac{\sin 2x}{e^x - 1} = \lim_{x \to 0} \frac{\sin 2x}{2x} \frac{2x}{e^x - 1} = 2$$

where we have also used the properties of limits of the Proposition 2.2.1.

2. We now want to compute the limit:

$$\lim_{x \to 0} \frac{1 - \cos x}{x^2} = \frac{0}{0}$$

We multiply the numerator and denominator by $1 + \cos x$, so to obtain the standard form as in (1) of Proposition 2.2.4.

$$\lim_{x \to 0} \frac{(1 - \cos x)(1 + \cos x)}{x^2(1 + \cos x)} = \lim_{x \to 0} \frac{(1 - \cos^2 x)}{x^2(1 + \cos x)} =$$

$$= \lim_{x \to 0} \frac{\sin^2 x}{x^2(1 + \cos x)} = \frac{1}{2}$$

where, once again, we used the properties in the Proposition 2.2.1.

2.3 INDETERMINATE FORMS

We want to compute some limits that appear in the indeterminate forms:

$$0/0 \qquad \infty/\infty \qquad +\infty - \infty.$$

In other words, limits of quotients (or differences) of functions that tend to zero or $\pm\infty$ for x approaching a point.

Definition 2.3.1 Let $f : D \longrightarrow \mathbb{R}$ be a function. We say that f is an *infinite* for x approaching c if:

$$\lim_{x \to c} f(x) = \pm\infty$$

Notice that we consider both cases $c = x_0$, with x_0 in $\mathbb{R}$, or $c = \pm\infty$. Also, we can define f infinite for x approaching c, using the right-hand or left-hand limit only.

Let us see some examples.

Example 2.3.2   1. The function $f(x) = x^n$ is an infinite for x approaching infinity (positive or negative). In fact, we have:

$$\lim_{x \to +\infty} x^n = +\infty, \qquad \lim_{x \to -\infty} x^n = (-1)^n \infty \quad n \text{ positive integer}$$

We leave to the reader, as an easy exercise, the proof of these statements.

2. The exponential and logarithmic function are both infinite for x that goes to $+\infty$. In fact, we have:

$$\lim_{x \to +\infty} e^x = +\infty, \qquad \lim_{x \to +\infty} \log(x) = +\infty$$

The logarithmic function is also an infinite when we take the right-hand limit:

$$\lim_{x \to 0^+} \log(x) = -\infty$$

We leave these statements to the reader as an exercise. Also, using a graphical software, it is easy to see geometrically that the exponential function and the logarithmic function take larger and larger values as the variable x increases. Similarly, we can look at the right-hand limit of the logarithmic function, for x approaching zero and justify the above equality from a graphical point of view.

Definition 2.3.3 Let us consider a function $f : D \longrightarrow \mathbb{R}$. We say f has *order of infinity* n, for x approaching c, with n positive integer, if:

1. $\lim_{x \to c} f(x) = \pm\infty$;

2. $\lim_{x \to c} \frac{f(x)}{x^n}$ exists and it is a non zero number.

We denote with $\text{ord}_\infty(f)$ the order of infinity of f for x approaching c.

Let us now consider two functions $f, g : D \longrightarrow \mathbb{R}$, which are infinite for x approaching c. We say that f has an order of infinity greater than the order of g, for x approaching c, if:

$$\lim_{x \to c} \frac{f(x)}{g(x)} = \pm\infty$$

We say that f has an order of infinity less than the order of g, for x approaching c, if:

$$\lim_{x \to c} \frac{f(x)}{g(x)} = 0$$

f and g have the same order of infinity if:

$$\lim_{x \to c} \frac{f(x)}{g(x)} = L \in \mathbb{R}, \quad L \neq 0$$

Observation 2.3.4 Let $c = +\infty$. We have that:

1. $\text{ord}_\infty(x^n) > \text{ord}_\infty(x^m)$, if $n > m$.

2. $\text{ord}_\infty(e^x) > \text{ord}_\infty(x^m)$, $\forall m$.

3. $\text{ord}_\infty(\log(x)) < \text{ord}_\infty(x^m)$, $\forall m$.

The first of these inequalities is seen graphically in the figure below. We invite the student to make use of a graphic software to check geometrically all these inequalities.

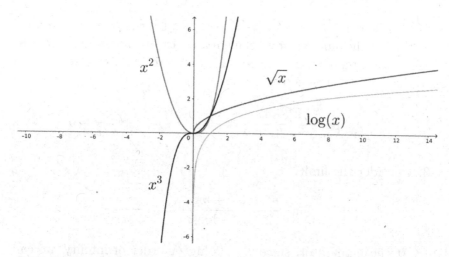

We notice also, by looking at the picture that n and m may also take rational values and all the statements are still true. Of course this is not a mathematical proof, but it gives a geometric intuition to comprehend our statements.

The following proposition is very important to compute limits involving indeterminate forms.

Proposition 2.3.5 *Let $f, g : D \longrightarrow \mathbb{R}$ be two functions, with f, g infinite for x approaching c. If $\operatorname{ord}_\infty(f) > \operatorname{ord}_\infty(g)$, then:*

$$\lim_{x \to c} f(x) + g(x) = \lim_{x \to c} f(x)$$

In other words: in a sum of infinites, we can ignore the lower-order infinite functions.

Proof. We can write:

$$\lim_{x \to c} f(x) + g(x) = \lim_{x \to c} f(x)[1 + g(x)/f(x)] = \lim_{x \to c} f(x)$$

□

Now we see some examples of applications of the notion of order of infinity.

Example 2.3.6   1. Consider $\lim_{x \to -\infty} x^3 - 4x = -\infty + \infty$. By the previous proposition, with $f(x) = x^3$ and $g(x) = -4x$, we can ignore $g(x)$, hence:

$$\lim_{x \to -\infty} x^3 - 4x = \lim_{x \to -\infty} x^3 = -\infty$$

2. Consider $\lim_{x\to-\infty} \frac{x^3-4x}{x^4-3x+2} = -\infty/\infty$. By the previous proposition, at the numerator we can ignore $-4x$, and at the denominator we can ignore $-3x+2$, because they are infinites of order less than x^3 e x^4 respectively. Hence:

$$\lim_{x\to-\infty} \frac{x^3-4x}{x^4-3x+2} = \lim_{x\to-\infty} \frac{x^3}{x^4} = \lim_{x\to-\infty} \frac{1}{x} = 0$$

3. Consider the limit:

$$\lim_{x\to+\infty} \frac{e^x + \sqrt{x} + \log(x)}{2+x^3-\log(x)} = \frac{\infty}{\infty-\infty}$$

At the numerator, since e^x has higher order of infinity, we can ignore the other terms. In the denominator, the higher order infinite is x^3, so we can ignore the other terms.

$$\lim_{x\to+\infty} \frac{e^x + \sqrt{x} + \log(x)}{2+x^3-\log(x)} = \lim_{x\to+\infty} \frac{e^x}{x^3} = +\infty$$

The last equality is due again to the fact that e^x has higher order of infinity.

We now proceed to describe a method for solving indeterminate forms of the type $\frac{0}{0}$.

Definition 2.3.7 Let us consider a function $f : D \longrightarrow \mathbb{R}$. We say f is has *infinitesimal order* n for x approaching c, with n positive integer, if:

1. $\lim_{x\to c} f(x) = 0$;

2. $\lim_{x\to c} \frac{f(x)}{x^n}$ exists and it is a non zero number.

Denote with $\mathrm{ord}_0(f)$ the order of infinitesimal of f at a given point.

Let us now consider two functions $f, g : D \longrightarrow \mathbb{R}$. We say that f has an infinitesimal order greater than g, for x approaching c, if:

$$\lim_{x\to c} \frac{f(x)}{g(x)} = 0$$

We say that f has an infinitesimal order smaller than g, for x approaching c, if:

$$\lim_{x \to c} \frac{f(x)}{g(x)} = \pm\infty$$

f and g have the same infinitesimal order if:

$$\lim_{x \to c} \frac{f(x)}{g(x)} = L \in \mathbb{R}, \quad L \neq 0$$

As before, we can also define the infinitesimal order for right and left-hand limits. As for the infinity case, we can observe graphically the following statements.

Observation 2.3.8 Let $c = 0$. We have that:

1. $\mathrm{ord}_0(x^n) > \mathrm{ord}_0(x^m)$, if $n > m$.

2. $\mathrm{ord}_0(e^{-1/x}) > \mathrm{ord}_0(x^m)$, $\forall m$ (right-hand limit).

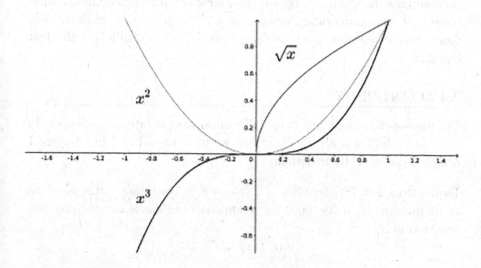

Proposition 2.3.9 *Let $f, g : D \longrightarrow \mathbb{R}$ be two functions. If $\mathrm{ord}_0(f) < \mathrm{ord}_0(g)$ then:*

$$\lim_{x \to c} f(x) + g(x) = \lim_{x \to c} f(x)$$

In other words: in a sum of functions, we can ignore the function with higher infinitesimal order (note the difference with respect to the previous case).

Proof. We have:

$$\lim_{x \to c} f(x) + g(x) = \lim_{x \to c} f(x)[1 + g(x)/f(x)] = \lim_{x \to c} f(x)$$

$\square$

This observation is important when calculating limits with indeterminate forms of the type $\frac{0}{0}$.

Example 2.3.10 We want to compute the limit:

$$\lim_{x \to 0^+} \frac{e^{-1/x} + \sqrt{x^3} + \log(1+x)}{x^3 - \log(1+2x)} = \lim_{x \to 0^+} \frac{\log(1+x)}{-\log(1+2x)} =$$

$$= \lim_{x \to 0^+} \frac{\log(1+x)}{x} \cdot \frac{x}{-\log(1+2x)} = -1/2$$

To compute this limit, we ignore the function with higher infinitesimal order. At the numerator we ignore $e^{-1/x}$ and $\sqrt{x^3}$, while at the denominator we ignore x^3. We use then standard limits for the last equality.

2.4 CONTINUITY

Continuous functions are extremely important in applied sciences; in fact, as we will see, all the functions we have introduced in Chapter 1 are continuous in their domain.

Definition 2.4.1 A function $f : D \longrightarrow \mathbb{R}$ is *continuous at a point* x_0 of its domain D, if its limit, for x approaching x_0, is the value of the function at x_0:

$$\lim_{x \to x_0} f(x) = f(x_0)$$

We say f is *continuous*, if it is continuous at all points in its domain.

In the previous section, we proved that:

$$\lim_{x \to x_0} x = x_0$$

Together with the properties of limits (Proposition 2.2.1), this allows us to say immediately that the polynomial, rational and algebraic functions are continuous.

We also have the following proposition.

Proposition 2.4.2 *The trigonometric functions* $\sin(x)$, $\cos(x)$, *the exponential and logarithmic functions* e^x, $\log(x)$ *are continuous.*

Proof. First, we prove the function $\sin(x)$ is continuous at $x = 0$:

$$\lim_{x \to 0} \sin(x) = \sin(0) = 0$$

Indeed

$$\lim_{x \to 0} \sin(x) = \lim_{x \to 0} \frac{\sin(x)}{x} x = \lim_{x \to 0} \frac{\sin(x)}{x} \lim_{x \to 0} x = 0$$

due to the standard limits. Similarly we have:

$$\lim_{x \to 0} \cos(x) = \lim_{x \to 0} \sqrt{1 - \sin^2(x)} = 1 = \cos(0)$$

because of the properties of limits.

Now we prove that the function $\sin(x)$ is continuous at a generic point x_0:

$$\lim_{h \to 0} \sin(x_0 + h) = \sin(x_0)$$

We use the formula:

$$\sin(x_0 + h) = \sin(x_0) \cos(h) + \cos(x_0) \sin(h)$$

Hence:

$$\lim_{h \to 0} \sin(x_0 + h) = \lim_{h \to 0} [\sin(x_0) \cos(h) + \cos(x_0) \sin(h)] = \sin(x_0)$$

We can prove the continuity of the function $\cos(x)$ in a similar way.

Let us now show that the exponential function is continuous at $x = 0$:

$$\lim_{x \to 0} e^x = 1$$

Reasoning as above:

$$\lim_{x \to 0} e^x - 1 = \lim_{x \to 0} \frac{e^x - 1}{x} x = \lim_{x \to 0} \frac{e^x - 1}{x} \lim_{x \to 0} x = 0$$

due to the standard limits. At the generic point x_0, we have:

$$\lim_{h \to 0} e^{x_0+h} - e^{x_0} = e^{x_0} \lim_{h \to 0} \frac{e^h - 1}{h} h = 0$$

due to the standard limits.

We leave the proof that the logarithmic function is continuous as an exercise. $\square$

Because of the properties of limits (Proposition 2.2.1), we can immediately state the following proposition.

Proposition 2.4.3 *Let $f, g : D \longrightarrow \mathbb{R}$ be continuous functions. Then*

1. *the sum $f + g$ is a continuous function.*

2. *the product $f \cdot g$ is a continuous function.*

3. *the quotient f/g is a continuous function for any x with $g(x) \neq 0$.*

4. *If f is continuous, then $f^{p/q}$ is continuous (in its domain), p, q positive integers.*

We also have the following proposition, whose proof we leave as an exercise.

Proposition 2.4.4 *The composition of continuous functions is a continuous function, that is, for f and g continuous functions, $f \circ g$ is continuous.*

Propositions 2.4.2, 2.4.3 and 2.4.4 allow us to determine the continuity of all the functions of interest in the applied sciences.

Example 2.4.5 We want to use the previous propositions to establish if the function

$$f(x) = \frac{e^{x^2+1} + \sin(2x)}{\log(x)}$$

is continuous. First, we determine its domain: $D = \{x \in \mathbb{R} | x > 0, x \neq 1\}$. Notice that the numerator is continuous as it is the sum of continuous functions: e^{x^2+1} and $\sin(2x)$ (they both are compositions of continuous functions). The denominator is also a continuous function. So, by Propositions 2.4.3, 2.4.4, the given function is continuous.

There are also functions which are not continuous: for example the step function of Example 2.1.3. This function is not continuous at the point $x = 0$, as there is no limit of the function at that point.

Let us also see an example where the limit at a given point exists, but the function is not continuous.

Example 2.4.6 Consider the function $f : \mathbb{R} \longrightarrow \mathbb{R}$:

$$f(x) = \begin{cases} 0 & x < 0 \\ 1 & x = 0 \\ 0 & x > 0 \end{cases}$$

We can readily see that:

$$\lim_{x \to 0^-} f(x) = 0 = \lim_{x \to 0^+} f(x)$$

Hence the limit for x approaching 0 of the function f exists and it is equal to zero. However, $f(0) = 1 \neq 0$, hence, even if the limit of the function f exists at $x = 0$, the function is not continuous.

Once again, we notice the following important fact: when calculating the limit of f for x approaching x_0, we are not interested in the value of f at the point x_0, but only in the values of f at points arbitrarily close to x_0.

2.5 DERIVATIVE OF A FUNCTION

The concept of derivative is the cornerstone of differential calculus and extremely important in all applications. Geometrically, the derivative $f'(x_0)$ of a function f at x_0 is the slope of the tangent line to the graph of f. That is, the derivative is the slope of the tangent line to the curve $y = f(x)$, at the point with x coordinate x_0. We obtain the derivative, by calculating the limit:

$$f'(x_0) = \lim_{h \to 0} \frac{f(x_0 + h) - f(x_0)}{h}$$

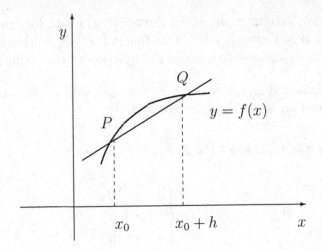

Indeed, from the picture we see that:

$$\tan \alpha = \frac{f(x_0 + h) - f(x_0)}{h}$$

represents the slope of the secant line through the points P and Q (recall: α is the angle formed by the line through P and Q and the x-axis). The quantity:

$$\frac{f(x_0 + h) - f(x_0)}{h}$$

is called the *difference quotient*. So, the derivative of f at the point x_0 is the limit for h approaching 0 of the difference quotient. Geometrically the derivative is then, as we anticipated, the slope of the tangent line to the curve $y = f(x)$ at the point $(x_0, f(x_0))$. Naturally, this is an intuitive and geometric reasoning, where we are assuming that the graph $y = f(x)$ describes a curve that admits a tangent line at all points.

We now give the rigorous definition.

Definition 2.5.1 Let $f : D \longrightarrow \mathbb{R}$ and x_0 a point in D. The derivative $f'(x_0)$ of f at x_0 is the limit (if it exists):

$$f'(x_0) = \lim_{h \to 0} \frac{f(x_0 + h) - f(x_0)}{h} \qquad (2.2)$$

If such a limit exists, we say that f is *differentiable* at the point x_0. If f is differentiable at every point of its domain D, we say that f is differentiable.

Let D' be the set of values in D for which we can compute the derivative of f. We can then define a function $f' : D' \longrightarrow \mathbb{R}$, that we call the *derivative* of f. We will denote the derivative f' of a function f also with the symbols:

$$\frac{df}{dx}, \quad D(f)$$

We notice that the definition of derivative $f'(x_0)$ in (2.2) can also be equivalently expressed as:

$$f'(x_0) = \lim_{x \to x_0} \frac{f(x) - f(x_0)}{x - x_0} \qquad (2.3)$$

where we set $x = x_0 + h$.

From the definition of derivative and differentiable function, we can immediately write the equation of the tangent line to the curve $y = f(x)$ at the point $(x_0, f(x_0))$:

$$y - f(x_0) = f'(x_0)(x - x_0)$$

Because of the properties of the limits in Proposition 2.2.1, we have the following result.

Proposition 2.5.2 *Let $f, g : D \longrightarrow \mathbb{R}$ be two differentiable functions. Then:*

1. $(f(x) + g(x))' = f'(x) + g'(x)$

2. $(kf(x))' = kf'(x)$

3. $k' = 0$

where $k \in \mathbb{R}$ is a constant and with k' we mean the derivative of the constant function.

The derivative of the product or the quotient of functions, is not as straightforward. In fact, to compute it, we need the *Leibniz rule*, that we state in the next proposition.

Proposition 2.5.3 *Let $f, g : D \longrightarrow \mathbb{R}$ be two differentiable functions. Then:*

1. Leibniz Rule:

$$(f(x)g(x))' = f'(x)g(x) + f(x)g'(x)$$

2. Derivative of the inverse of a function:

$$D(\frac{1}{f(x)}) = -\frac{f'(x)}{f(x)^2}, \qquad f(x) \neq 0$$

3. Derivative of the quotient of two functions:

$$D(\frac{f(x)}{g(x)}) = \frac{f'(x)g(x) - f(x)g'(x)}{g(x)^2}, \qquad g(x) \neq 0$$

Proof. Let $F(x) := f(x)g(x)$. We write the difference quotient:

$$\frac{F(x+h)-F(x)}{h} = \frac{f(x+h)g(x+h)-f(x)g(x)}{h} =$$

$$= \frac{f(x+h)g(x+h)-f(x)g(x+h)+f(x)g(x+h)-f(x)g(x)}{h} =$$

$$= \frac{(f(x+h)-f(x))g(x+h)+f(x)(g(x+h)-g(x))}{h} =$$

$$= \frac{(f(x+h)-f(x))}{h}g(x+h) + f(x)\frac{(g(x+h)-g(x))}{h}$$

Taking the limit for h approaching zero, we obtain the Leibniz rule.

(2) Since the derivative of a constant is zero, applying (1) we have:

$$0 = \left(f(x) \cdot \frac{1}{f(x)} \right)' = f'(x)\frac{1}{f(x)} + f(x) \left(\frac{1}{f(x)} \right)'$$

and we obtain (2).

(3) is a consequence of (1) and (2), applying the Leibniz rule to the product of f and $1/g$. $\qquad \square$

This proposition allows us to compute immediately the derivatives of polynomial and rational functions.

Example 2.5.4   1. We compute the derivative of the function $f(x) = x^n$. We see that for $n = 1$ we can compute the derivative of x directly from the definition:

$$\lim_{h \to 0} \frac{(x+h) - x}{h} = \lim_{h \to 0} \frac{h}{h} = 1$$

We see now the case of x^n. Let us repeat n times the Leibniz Rule. We have:

$$(x^n)' = (x^{n-1}x)' = (x^{n-1})'x + x^{n-1} =$$

$$= (x^{n-2})'x^2 + x^{n-1} + x^{n-1} = \ldots$$

$$= (x)'x^{n-1} + x^{n-1} + \cdots + x^{n-1} = nx^{n-1}$$

2. We now compute the derivative of a polynomial function: $f(x) = 2x^5 - 3x^2 + 4$. Since the derivative of a sum is the sum of the derivatives we have:

$$f'(x) = (2x^5)' - (3x^2)' + (4)' = 10x^4 - 6x$$

3. Let us compute the derivative of a rational function:

$$D\left(\frac{2x^4 - x^3}{x^2 - 1}\right) = \frac{(2x^4 - x^3)'(x^2 - 1) - (2x^4 - x^3)(x^2 - 1)'}{(x^2 - 1)^2} =$$

$$= \frac{(8x^3 - 3x^2)(x^2 - 1) - (2x^4 - x^3)(2x)}{(x^2 - 1)^2} =$$

$$= \frac{8x^5 - 3x^4 - 8x^3 + 3x^2 - 4x^5 + 2x^4}{(x-1)^2(x+1)^2} =$$

$$= \frac{4x^5 - x^4 - 8x^3 + 3x^2}{(x-1)^2(x+1)^2}$$

Let us now look at a proposition giving the derivatives of important functions in applied sciences.

Proposition 2.5.5 *The exponential, logarithmic, sine and cosine functions are differentiable in their domain and we have:*

1. $(e^x)' = e^x$

2. $(\log(x))' = 1/x$

3. $(\sin(x))' = \cos(x)$

4. $(\cos(x))' = -\sin(x)$.

Proof. (1) Compute the limit of the difference quotient:

$$\lim_{h \to 0} \frac{e^{x+h} - e^x}{h} = e^x \lim_{h \to 0} \frac{e^h - 1}{h} = e^x$$

using one of the standard limits.

We can prove in a similar way all the other properties, using standard limits. ☐

We conclude the section with the chain rule, that will allow us to take the derivative of the composition of two functions. Recall that, given the functions f and g, their composition $f \circ g$ is the function:

$$(f \circ g)(x) = f(g(x))$$

Proposition 2.5.6 Chain rule: *Let f and g be differentiable functions and $f \circ g$ their composition. Then:*

$$(f \circ g)'(x) = f'(g(x))g'(x)$$

Proof. We prove the statement assuming there is an open interval containing x_0, where $g(x) \neq 0$. This is a special case; however, this assumption is satisfied for most functions of interest in applied sciences. The proof in the general case is a small variation of this argument.

By definition of derivative, we have:

$$(f \circ g)'(x_0) = \lim_{x \to x_0} \frac{f(g(x)) - f(g(x_0))}{x - x_0} =$$

$$= \lim_{x \to x_0} \frac{f(g(x)) - f(g(x_0))}{g(x) - g(x_0)} \cdot \frac{g(x) - g(x_0)}{x - x_0} = f'(g(x_0))g'(x_0)$$

☐

We now see some examples of application of the chain rule.

Example 2.5.7 1. We want to compute the derivative of the function e^{2x^2+1}. In this case $f(x) = e^x$, $g(x) = 2x^2 + 1$, from which $f'(x) = e^x$ and $g'(x) = 4x$. Hence:

$$D(e^{2x^2+1}) = e^{2x^2+1} \cdot 4x$$

2. We want to compute the derivative of the function $\sqrt{\sin(x^2 - 7)}$. It is the composition of three functions:

$$f(x) = \sqrt{x}, \qquad g(x) = \sin(x), \qquad h(x) = x^2 - 7$$

We apply the chain rule twice:

$$D(\sqrt{\sin(x^2 - 7)}) = f'(g(h(x))g'(h(x))h'(x) =$$

$$= \tfrac{1}{2}[\sin(x^2 - 7)]^{-1/2} \cos(x^2 - 7)\, 2x =$$

$$= \frac{x \cos(x^2 - 7)}{\sqrt{\sin(x^2 - 7)}}$$

2.6 DERIVABILITY AND CONTINUITY

In the previous sections we introduced the concepts of continuous and differentiable function. Now we want to understand how these notions are related with each other. We begin with a proposition establishing that every differentiable function is indeed continuous.

Proposition 2.6.1 *Let $f : D \longrightarrow \mathbb{R}$ be a differentiable function at a point x_0 of its domain. Then f is continuous at x_0.*

Proof. We have to check that:

$$\lim_{x \to x_0} f(x) = f(x_0)$$

or equivalently:

$$\lim_{x \to x_0} f(x) - f(x_0) = 0$$

We multiply and divide $f(x) - f(x_0)$ by $x - x_0$:

$$\lim_{x \to x_0} \frac{f(x) - f(x_0)}{x - x_0} \cdot (x - x_0) = f'(x_0) \cdot 0 = 0$$

□

Let us see now, with a counterexample, that not all continuous functions are differentiable.

Example 2.6.2 Consider the function $f(x) = |x|$, that is:

$$f(x) = \begin{cases} -x & x < 0 \\ x & x \geq 0 \end{cases}$$

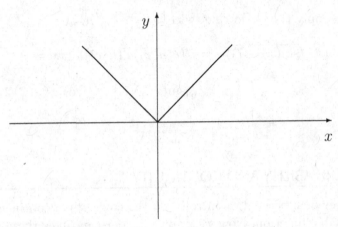

This function is continuous, in fact both functions $f_1(x) = -x$ and $f_2(x) = x$ are continuous at all x. So the only point, we need to worry about, is the point $x_0 = 0$. We compute the right and left-hand limits:

$$\lim_{x \to 0^-} f(x) = \lim_{x \to 0^-} -x = 0, \qquad \lim_{x \to 0^+} f(x) = \lim_{x \to 0^+} x = 0$$

Since these limits are equal, we have that the limit of f for x approaching zero exists and it is equal to 0, which is the value of the f function at the point $x = 0$. Hence f is continuous, since it is continuous for any real x.

We now see that f is not differentiable at $x = 0$. We compute left and right-hand limits of the difference quotient:

$$\lim_{x \to 0^-} \frac{f(x) - f(0)}{x - 0} = \frac{-x}{x} = -1, \qquad \lim_{x \to 0^+} \frac{f(x) - f(0)}{x - 0} = \frac{x}{x} = 1.$$

Since the right and left-hand limits of the difference quotient, for x approaching zero, do not coincide, we have that the derivative of f does not exist at the point $x = 0$. Hence, we proved that f is continuous, but not differentiable, at the point $x = 0$. Looking at the above figure, this is geometrically clear: we cannot draw the tangent line at the point $x = 0$ to the graph of f.

2.7   DE L'HOPITAL'S RULE

The method of De L'Hopital (or De L'Hospital) allows us to compute limits leading to the indeterminate forms $0/0$ or ∞/∞ through the derivative. We will state this result in the particular case, of great interest, of continuous, differentiable functions, with continuous derivative. We could prove this result also without such hypotheses, but since most functions we are interested in satisfy them, we will not examine the more general case.

Theorem 2.7.1 De L'Hopital's Rule: *Let f , $g : [a, b] \longrightarrow \mathbb{R}$ be continuous functions in $[a, b]$, differentiable in (a, b) except, eventually, in $c \in (a, b)$. Let $g'(x) \neq 0$, for $x \neq c$ and suppose:*

$$\lim_{x \to c} f(x) = \lim_{x \to c} g(x) = 0$$

or

$$\lim_{x \to c} f(x) = \pm\infty, \qquad \lim_{x \to c} g(x) = \pm\infty$$

Moreover assume:

$$\lim_{x \to c} \frac{f'(x)}{g'(x)} = L$$

Then

$$\lim_{x \to c} \frac{f(x)}{g(x)} = L$$

(Notice L can also be $\pm\infty$).

Proof. We prove the result only for the case $c \in \mathbb{R}$, that is, the values of the functions f, g approaching zero, as x approaches c. The general case is a variation of the argument given below. We have that

$$\lim_{x \to c} \frac{f(x)}{g(x)} = \lim_{x \to c} \frac{f(x) - 0}{g(x) - 0} = \lim_{x \to c} \frac{f(x) - f(c)}{g(x) - g(c)}$$

$$= \lim_{x \to c} \frac{\left(\frac{f(x) - f(c)}{x - c}\right)}{\left(\frac{g(x) - g(c)}{x - c}\right)} = \frac{\lim_{x \to c}\left(\frac{f(x) - f(c)}{x - c}\right)}{\lim_{x \to c}\left(\frac{g(x) - g(c)}{x - c}\right)} = \frac{f'(x)}{g'(x)}$$

$\square$

Observation 2.7.2 At this point, the student may be tempted to prove the standard limits in Proposition 2.2.4 using De L'Hospital's rule, but this would be a serious logical error. Let us see why. Suppose we want to prove the standard limit:

$$\lim_{x \to 0} \frac{e^x - 1}{x}$$

using De L'Hospital's rule. Let us take the derivative of the numerator and the denominator:

$$\lim_{x \to 0} \frac{e^x - 1}{x} = \lim_{x \to 0} \frac{e^x}{1} = 1$$

The calculation is correct, but the logical reasoning is wrong! Indeed, in order to prove that the derivative of e^x is equal to e^x we used precisely this standard limit.

Hence, De L'Hospital's rule is very effective to compute limits; however, we must pay close attention, if we want to use it to prove results.

2.8   DERIVATIVE OF THE INVERSE FUNCTION

We want to compute the derivative of the inverse function of a given function. In other words, we want to compute the derivative of f^{-1}, if we know the derivative of f.

We recall that, given a function $f : D \longrightarrow \mathbb{R}$, its inverse (if it exists) is a function, which we denote by f^{-1}, which has the following property:

$$f(f^{-1}(x)) = x \qquad f^{-1}(f(x)) = x.$$

We have already encountered some examples in the first chapter: the inverse of the exponential function is the logarithmic function (taking into account the domain) and we saw some important trigonometric inverse functions: arcsine, arccosine and arctangent.

To compute the derivative of the inverse function, we use the chain rule:

$$D(f(f^{-1}(x))) = f'(f^{-1}(x))(f^{-1})'(x) = 1 \implies (f^{-1})'(x) = \frac{1}{f'(f^{-1}(x))}$$
$$(2.4)$$

since $f(f^{-1}(x)) = x$ and since the derivative of the identity function is 1 and assuming that:

$$f'(f^{-1}(x)) \neq 0$$

Let us see, how to compute the derivative of the arctangent function (see Chapter 1, for the definition). The derivatives of the other inverses of trigonometric functions are computed in a similar way (see the Appendix 2.12).

Example 2.8.1 We want to prove that:

$$\frac{d}{dx} \arctan(x) = \frac{1}{1 + x^2}$$

In order to apply more easily the formula (2.4), we set $y = f^{-1}(x)$. Then, the formula (2.4) becomes:

$$(f^{-1})'(x) = \frac{1}{f'(y)} \tag{2.5}$$

We compute the derivative of the function $f(y) = \tan(y)$, using the quotient rule:

$$\frac{d}{dy} \tan(y) = \frac{d}{dy} \frac{\sin(y)}{\cos(y)} = \frac{\cos(y)\cos(y) + \sin(y)\sin(y)}{\cos(y)^2} = \frac{1}{\cos^2(y)}$$

From the formula (2.5):

$$\frac{d}{dx} \arctan(x) = \frac{1}{\frac{d}{dy} \tan(y)}$$

Hence

$$\frac{d}{dx} \arctan(x) = \frac{1}{\frac{1}{\cos^2(y)}} = \cos^2(y) = \frac{1}{1 + \tan^2(y)} \tag{2.6}$$

the last equality is a simple application of the definitions and of the Pythagorean trigonometric identity: $\sin^2(y) + \cos^2(y) = 1$.

Since we set $y = f^{-1}(x)$, we immediately have $f(y) = f(f^{-1}(x)) = x$. Hence, $x = \tan(y)$ and substituting into the equation (2.6) we have:

$$\frac{d}{dx} \arctan(x) = \frac{1}{1 + x^2}$$

2.9 EXERCISES WITH SOLUTIONS

2.9.1 Using the definition of limit, check that:

$$\lim_{x \to 0^+} 3x - 2 = -2$$

Solution. Given $\epsilon > 0$, we have to find $\delta > 0$ such that, if $0 < x < \delta$, we have $|3x - 2 - (-2)| < \epsilon$, that is: $3x < \epsilon$ for $x > 0$ (the only case that interests us). Hence, if $\delta = \epsilon/3$, we get the desired inequality.

2.9.2 Using the definition of limit, check that:

$$\lim_{x \to +\infty} 1/x^2 = 0$$

Solution. We must show that, for every $\epsilon > 0$, there exists $M > 0$, such that for every $x \in \mathbb{R}$, with $x > M$, we have $|1/x^2| < \epsilon$, i.e. $1/x^2 < \epsilon$. From the latter inequality, we get:

$$x^2 - 1/\epsilon > 0$$

which is satisfied for $x < -1/\sqrt{\epsilon}$, $x > 1/\sqrt{\epsilon}$, (recall ϵ is a positive number). So, taking $M = 1/\sqrt{\epsilon}$, we get that

$$|f(x) - L| = |1/x^2 - 0| < \epsilon$$

as requested.

2.9.3 Compute the limit:

$$\lim_{x \to 0^+} \frac{1}{x}(\log(1 + 2x) + e^{-1/x}x)$$

Solution. It is an indeterminate form $\frac{0}{0}$. We observe that we can write this limit as:

$$\lim_{x \to 0^+} \frac{\log(1 + 2x) + e^{-1/x}}{x}$$

In the numerator, we have $\mathrm{ord}_0(e^{-1/x}) > \mathrm{ord}_0(\log(1 + 2x))$, since $e^{-1/x}$ has infinitesimal order greater than any power of x (check this by exercise or with a graphic calculator geometrically).

Hence, we can ignore $e^{-1/x}$ and write:

$$\lim_{x \to 0^+} \frac{\log(1 + 2x)}{x} = \lim_{x \to 0^+} 2\frac{\log(1 + 2x)}{2x} = 2$$

because of standard limits.

2.9.4 Compute the limit:

$$\lim_{x \to +\infty} \frac{e^{-x}}{\sin(1/x) + x^{-2}}$$

Solution. This is the indeterminate form $\frac{0}{0}$. We use infinitesimal order. By the standard limits, we see that $\sin(1/x)$ has infinitesimal order equal to $1/x$, hence less than the infinitesimal order of $1/x^2$, as x approaches infinity. We can then ignore this last term:

$$\lim_{x \to +\infty} \frac{e^{-x}}{\sin(1/x) + x^{-2}} = \lim_{x \to +\infty} \frac{e^{-x}}{\sin(1/x)} = 0$$

This is because e^{-x} has infinitesimal order smaller than any power of $1/x$, as x approaches infinity, while $\sin(1/x)$ has order of infinitesimal equal to $1/x$.

We invite the student to try to solve this limit through De L'Hospital's rule: the calculation turns out to be much more complicated.

2.9.5 Compute the limit:

$$\lim_{x \to +\infty} \frac{e^x + x^2}{\sin(x) + \log(x^2 - 5)}$$

Solution. This is the indeterminate form $\frac{\infty}{\infty}$. We notice that even if the limit for x approaching infinity of $\sin(x)$ does not exist, we can ignore this term because it is bounded and between -1 and 1, since $\log(x^2 - 5)$ tends to infinity.

Let us proceed with De L'Hospital's rule:

$$\lim_{x \to +\infty} \frac{e^x + x^2}{\log(x^2 - 5)} = \lim_{x \to +\infty} \frac{e^x + 2x}{\frac{2x}{x^2 - 5}} = \lim_{x \to +\infty} \frac{(e^x + 2x)(x^2 - 5)}{2x}$$

Repeating again the rule:

$$\lim_{x \to +\infty} \frac{(e^x + 2x)(x^2 - 5)}{2x} = \lim_{x \to +\infty} \frac{(e^x + 2)(x^2 - 5) + (e^x + 2x)(2x)}{2} = +\infty$$

where at the numerator we applied the Leibniz rule. We notice that we could have arrived at the same conclusion more quickly by noticing that e^x has order of infinity greater than all the functions appearing in this limit.

2.9.6 Determine the tangent line at the point $x_0 = 0$ to the curve $y = e^{\sin(x)}$.

Solution. We recall that the tangent line to graph of $f(x)$ at the point x_0 has equation:

$$y - f(x_0) = f'(x_0)(x - x_0)$$

The derivative of $f(x) = e^{\sin(x)}$ is given by $e^{\sin(x)}\cos(x)$, by the chain rule. Hence, $f'(0) = 1$ and $f(0) = 1$. Substituting into the previous expression:

$$y - 1 = 1(x - 0)$$

Hence the tangent line is given by the equation: $y = x + 1$.

2.10 SUGGESTED EXERCISES

2.10.1 Give clearly the following definitions of limit $(a \in \mathbb{R})$:

1. $\lim_{x \to a+} f(x) = +\infty$

2. $\lim_{x \to -\infty} f(x) = L$

3. $\lim_{x \to a-} f(x) = -\infty$

4. $\lim_{x \to +\infty} f(x) = -\infty$

2.10.2 Check, using the definition, the result of the following the limits:

1. $\lim_{x \to 2+} 1/(x - 2) = +\infty$

2. $\lim_{x \to 3} 3x - 4 = 5$

3. $\lim_{x \to -\infty} 2x^3 = -\infty$

4. $\lim_{x \to 0+} \log(x) = -\infty$

5. $\lim_{x \to +\infty} \log(x) = +\infty$

6. $\lim_{x \to +\infty} 1/x = 0$

7. $\lim_{x \to +\infty} e^x = +\infty$

8. $\lim_{x \to +\infty} e^{-1/x} = 0$

9. $\lim_{x \to +\infty} 2x - 3 = +\infty$

10. $\lim_{x \to 0} e^x = 1$

11. (*) $\lim_{x \to a} e^x = e^a$

12. (*) $\lim_{x \to -2} x^3 = -8$

2.10.3 Check, using the definition, the result of the following limits:

1. $\lim_{x \to +\infty} x^n = (-1)^n \infty$

2. $\lim_{x \to +\infty} x^n = +\infty$

2.10.4 Compute the following limits using the standard limits.

1. $\lim_{x \to 0+} \frac{\tan(x)}{x}$

2. $\lim_{x \to 0+} \frac{1 - \cos(x)}{x}$

3. $\lim_{x \to 1+} \frac{\log(x)}{1 - x}$

4. $\lim_{x \to 0} \frac{1 - \cos(x) - 2\sin(x)}{x}$

5. $\lim_{x \to 0} \frac{2 - \sin(x) - 2e^{3x}}{x}$

6. $\lim_{x \to 0} \frac{e^{3x} - 1}{\log(2x + 1)}$

2.10.5 Compute the following limits (when possible use two different methods):

1. $\lim_{x \to 0+} \frac{x^3 - x^2 + x}{x^5 + 6x}$

2. $\lim_{x \to -2+} \frac{x - 1}{x^2 - x - 6}$

3. $\lim_{x \to -\infty} \frac{x^3 - x^2 + x}{x^5 + 6x + 4}$

4. $\lim_{x \to +\infty} \frac{4x^3 - x^2 + x + 5}{5x^3 + 6x + 1}$

5. $\lim_{x \to 0+} \frac{\sin(2x) + x + x^4}{x^2 - 6x}$

6. $\lim_{x \to 0+} \frac{\cos(3x) - 1 + x^4 + x^2}{e^{x^2} - 1 + 2x^2 + 4x^3}$

7. $\lim_{x \to 0+} \frac{\log(1 + x^2) + x^4 + x^2}{\sqrt{1 + x^2} - 1}$

8. $\lim_{x\to 0+} \frac{\log(\cos x)}{\tan(x)\sin(x)}$

9. $\lim_{x\to 0+} \frac{\log(2-\cos(x))}{\sin^2(x)+2x^2+4x^3}$

10. $\lim_{x\to 0+} \frac{e^{\sin x}+3x+x^2}{\log(1+x)}$

11. $\lim_{x\to 0+} (\log(x) - \arctan(x-1))$

12. $\lim_{x\to+\infty} \sqrt{3}\,(x-1)(x-2)^2$

13. $\lim_{x\to+\infty} x^2 \left(\frac{1}{3x^2-1}\right)$

14. $\lim_{x\to 0+} \frac{\cos^2 x - 1 + x^2}{\sin^2 x + x^2}$

15. $\lim_{x\to 0+} \frac{(1+x)^6 - e^{\sin x}}{1-\cos(x^2+3x)}$

2.10.6 Compute the following limits (with any method):

1. $\lim_{x\to 0+} \frac{\sin(x)-x^2}{(\tan^2(x)+\log(1+x)}$

2. $\lim_{x\to 0+} \frac{(\cos(x)-1)-\log(x+1)-x^6}{e^{-1/x}+x^2\sin(x)}$

3. $\lim_{x\to 0+} \frac{(\cos(x)-1)^2+\log(x+1)^2-x^5}{e^{-1/x}+x\sin(x)}$

4. $\lim_{x\to 0+} \frac{\cos(x)-1+\log(x+1)-x^3}{e^{-1/x}+x\sin(x)}$

2.10.7 Compute the derivative of the following functions.

1. $(x^2 + 3x)e^{2x}$

2. $\sqrt{3x^3 - 5x + 1}$

3. $(-x^2 + 3x)^{1/3}$

4. $\frac{2}{1+3x^2}$

5. $\frac{x+2}{1+x^2}$

6. $\frac{\cos^2 x}{\sin^2 x + 3x^2}$

7. $e^{\sin 2x}$

8. $\log\left(\frac{x}{3+x^2}\right)$

9. $\tan(x^2 + 3x)$

10. $\frac{(1+x)}{e^{\cos x}}$

11. $x\log\left(\frac{1}{3x+1}\right)$

2.10.8 Write the equation for the tangent line at the point x_0 to the graphs of the following functions:

1. $(x^2 + 3x)e^{2x}$, $x_0 = 0$

2. $\sqrt{3x^3 - 5x + 1}$, $x_0 = 0$

3. $(-x^2 + 3x)^{1/3}$, $x_0 = 1$

4. $\frac{2}{1+3x^2}$, $x_0 = -1$

5. $\frac{x+2}{1+x^2}$, $x_0 = 0$

6. $e^{\sin 2x}$, $x_0 = 0$

7. $\tan(x^2 + 3x)$, $x_0 = 0$

2.10.9 Find, if possible, the points where the graph of $f(x)$ has horizontal tangent line:

1. $f(x) = \log(x^2 + 3)$

2. $f(x) = e^{\frac{1}{x+1}}$

2.10.10 Consider the statement: "Every continuous function is differentiable". If it is true, give a proof, if it is false, provide a counterexample.

2.10.11 Find, if possible, the points where the graph of the function $f(x)$ has horizontal tangent line.

$$f(x) = \log\frac{x+2}{x-3}$$

Compute also:

$$\lim_{x\to\infty} f(x) \qquad \lim_{x\to-\infty} f(x)$$

2.10.12 Assuming (1) and (2) of Proposition 2.5.3, prove the formula for the derivative of a quotient of functions:

$$\frac{f(x)}{g(x)} = \frac{f'(x)g(x) - f(x)g'(x)}{g(x)^2}, \qquad g(x) \neq 0$$

2.11   APPENDIX: DERIVATION RULES

We list the most common derivation rules that we have discussed in this chapter.

1. $\frac{d}{dx}(f+g) = \frac{df}{dx} + \frac{dg}{dx}$

2. $\frac{d}{dx}(c \cdot f) = c \cdot \frac{df}{dx}$

3. $\frac{d}{dx}(f \cdot g) = f \cdot \frac{dg}{dx} + g \cdot \frac{df}{dx}$

4. $\frac{d}{dx}\left(\frac{f}{g}\right) = \dfrac{-f \cdot \dfrac{dg}{dx} + g \cdot \dfrac{df}{dx}}{g^2}$

5. $\frac{d}{dx}[f(g(x))] = \frac{df}{dg} \cdot \frac{dg}{dx} = f'(g(x)) \cdot g'(x)$

6. $\frac{d}{dx}\left(\frac{1}{f}\right) = -\frac{f'}{f^2}$

2.12   APPENDIX: DERIVATIVES

We give the derivatives of the most common functions.

1. $\frac{d}{dx}(c) = 0, c \in \mathbb{R}$

2. $\frac{d}{dx}x = 1$

3. $\frac{d}{dx}x^n = nx^{n-1}$

4. $\frac{d}{dx}\sqrt{x} = \frac{1}{2\sqrt{x}}$

5. $\dfrac{d}{dx}\dfrac{1}{x} = -\dfrac{1}{x^2}$

6. $\dfrac{d}{dx}\sin(x) = \cos(x)$

7. $\dfrac{d}{dx}\cos(x) = -\sin(x)$

8. $\dfrac{d}{dx}\tan(x) = \dfrac{1}{\cos^2(x)}$

9. $\dfrac{d}{dx}e^x = e^x$

10. $\dfrac{d}{dx}a^x = a^x\log(a)$ if $a > 0$

11. $\dfrac{d}{dx}\log(x) = \dfrac{1}{x}$

12. $\dfrac{d}{dx}\log_a(x) = \dfrac{1}{x\log(a)}$ if $a > 0$, $a \neq 1$

13. $\dfrac{d}{dx}(f^g) = \dfrac{d}{dx}\left(e^{g\log(f)}\right) = f^g\left(f'\dfrac{g}{f} + g'\log(f)\right)$, $f > 0$

14. $\dfrac{d}{dx}(c^f) = \dfrac{d}{dx}\left(e^{f\log(c)}\right) = c^f\log(c)\cdot f'$

15. $\dfrac{d}{dx}\arcsin(x) = \dfrac{1}{\sqrt{1-x^2}}$

16. $\dfrac{d}{dx}\arccos(x) = -\dfrac{1}{\sqrt{1-x^2}}$

17. $\dfrac{d}{dx}\arctan(x) = \dfrac{1}{x^2+1}$

2.13 APPENDIX: THEOREMS ON LIMITS

In this appendix we prove two key results of the theory on limits: the Uniqueness of limit Theorem and the Comparison Theorem.

Theorem 2.13.1 Uniqueness of limit: *Let $f : D \to \mathbb{R}$ and suppose f is defined for all x in an open interval containing x_0, but not necessarily for x_0. If there is a limit of f as x approaches x_0, then this limit is unique.*

Proof. Suppose L_1 and L_2 are two limits of the function f as x approaches x_0. We will show that $L_1 = L_2$. By contradiction, we assume that $L_1 \neq L_2$. Then $\exists\, \epsilon_1,\ \epsilon_2 > 0$ such that

$$(L_1 - \epsilon_1, L_1 + \epsilon_1) \cap (L_2 - \epsilon_2, L_2 + \epsilon_2) = \emptyset.$$

By definition of limit $\exists\, \delta_1,\ \delta_2 > 0$, such that $\forall x \in (x_0 - \delta_1, x_0 + \delta_1)$,

$$f(x) \in (L_1 - \epsilon_1, L_1 + \epsilon_1)$$

and $\forall x \in (x_0 - \delta_2, x_0 + \delta_2)$,

$$f(x) \in (L_2 - \epsilon_2, L_2 + \epsilon_2).$$

Let $x \in (x_0 - \delta_1, x_0 + \delta_1) \cap (x_0 - \delta_2, x_0 + \delta_2) = (x_0 - \delta, x_0 + \delta)$ then

$$f(x) \in (L_1 - \epsilon_1, L_1 + \epsilon_1) \cap (L_2 - \epsilon_2, L_2 + \epsilon_2) = \emptyset$$

and this is a contradiction. □

We now state and prove the Comparison Theorem, also known as the Sandwich Theorem.

Theorem 2.13.2 Comparison theorem: *Consider the functions f, g, $h : D \to \mathbb{R}$, defined for all x in an open interval containing x_0, but not necessarily for x_0. If*

$$\lim_{x \to x_0} f(x) = L = \lim_{x \to x_0} h(x)$$

and if $\exists \delta > 0$ such that:

$$f(x) \leq g(x) \leq h(x),$$

for $\forall x \in (x_0 - \delta, x_0 + \delta) \setminus \{x_0\}$ then

$$\lim_{x \to x_0} g(x) = L$$

Proof. By the definition of limit $\forall \epsilon > 0$, $\exists\, \delta_1,\, \delta_2 > 0$ such that

$$L - \epsilon < f(x) < L + \epsilon$$

$\forall x \in (x_0 - \delta_1, x_0 + \delta_1) \setminus \{x_0\}$ and

$$L - \epsilon < h(x) < L + \epsilon$$

$\forall x \in (x_0 - \delta_2, x_0 + \delta_2) \setminus \{x_0\}$. Hence

$$L - \epsilon < f(x) \le g(x) \le h(x) < L + \epsilon$$

$\forall x \in (x_0 - \delta_1, x_0 + \delta_1) \cap (x_0 - \delta_2, x_0 + \delta_2) \cap (x_0 - \delta, x_0 + \delta) \setminus \{x_0\}$. Hence $\forall \epsilon > 0 \; \exists\, \delta' > 0$ such that
$\forall x \in (x_0 - \delta', x_0 + \delta') \setminus \{x_0\} = (x_0 - \delta_1, x_0 + \delta_1) \cap (x_0 - \delta_2, x_0 + \delta_2) \cap (x_0 - \delta, x_0 + \delta) \setminus \{x_0\}$

$$L - \epsilon \le g(x) \le L + \epsilon$$

in other words

$$\lim_{x \to x_0} g(x) = L.$$

$\square$

CHAPTER 3

Applications of the Derivative

3.1 THE LINEAR APPROXIMATION

The concept of linear approximation is based on the geometric interpretation of the derivative. In fact, as we saw in Chapter 2, we can view the derivative of a function f at a point, as the slope of the tangent line to the curve $y = f(x)$ at that point. Consider a function $f : D \longrightarrow \mathbb{R}$ and two points of the domain P and Q, with coordinates $(x_0, f(x_0))$ and $(x_0 + h, f(x_0 + h))$, respectively. We see that the line passing through the points P and Q approximates the tangent line to graph $y = f(x)$ and also the graph itself. In particular, we get a good approximation, by choosing h very small. We emphasize once again that this is just an intuitive reasoning: only the definition of limit can rigorously account for such notions such as "near", "approximate" and so on.

DOI: 10.1201/9781003343288-3

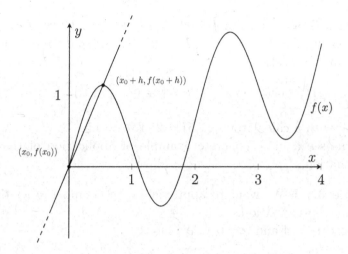

As we see from the graph, the difference quotient

$$\frac{f(x_0 + h) - f(x_0)}{h}$$

represents the slope of the line passing through P and Q. If we consider a "small" value of h, we can effectively approximate the derivative as

$$f'(x_0) \simeq \frac{f(x_0 + h) - f(x_0)}{h}$$

The symbol "$\simeq$" means that we have an approximation and not an equality.

We obtain an estimate for the value of $f(x_0 + h)$, which is called *linear approximation* of f:

$$f(x_0 + h) \simeq f(x_0) + f'(x_0) \cdot h. \qquad (3.1)$$

We can also express it more concisely as:

$$\Delta(f) \simeq f'(x_0)\Delta x, \quad \text{where} \quad \Delta(f) = f(x_0 + h) - f(x_0), \quad \Delta x = h.$$
$$(3.2)$$

For example, if we consider the function $f(x) = 2x^2$, for $x_0 = 1$ and $h = 0.02$, we can write the formula (3.1) as:

$$f(1.02) \cong f(1) + f'(1) \cdot 0.02 = 2 + 4 \cdot 1 \cdot 0.02 = 2.08$$

Checking with a calculator: $f(1.02) = 2 \cdot 1.02^2 = 2.0808$. If we choose a larger value of h, for example, $h = 0.5$, we see that we obtain a less effective approximation

$$f(1.5) \cong f(1) + f'(1) \cdot 0.5 = 2 + 4 \cdot 1 \cdot 0.5 = 4$$

Checking with a calculator: $f(1.5) = 2 \cdot 1.5^2 = 4.5$.

Let us see a more concrete example of application of the linear approximation.

Example 3.1.1 We want to approximate the number $\sqrt{4.1}$ through the linear approximation.

Choose $x_0 = 4$ and $h = 0.1$. We have:

$$\begin{aligned} f(4.1) &\simeq f(4) + 0.1 f'(4) = \\ &= 2 + \frac{0.1}{2\sqrt{4}} = 2.025 \end{aligned} \tag{3.3}$$

Checking with a calculator: $f(4.1) = \sqrt{4.1} = 2.02482$; hence, the linear approximation is a good approximation of the given number.

We now make an important observation, which explains better the geometric meaning of the linear approximation. It also justifies the choice of the term "linear approximation."

Observation 3.1.2 In Leibniz notation, we write the derivative of a function as:

$$\frac{df}{dx} = \frac{\text{variation of } f \text{ along the tangent line}}{\text{variation of } x}$$

In the linear approximation, we approximate the variation $\Delta(f) := f(x_0 + h) - f(x_0)$ of a function f, through its *differential* df, that is the variation $df = f'(x_0)(x - x_0)$. In fact, $y = f(x_0) + f'(x_0)(x - x_0)$ represents the *tangent line* to $y = f(x)$ at the point x_0. So, we have:

$$df = y - f(x_0) = f'(x_0)(x - x_0), \quad \text{while} \quad \Delta(f) = f(x_0 + h) - f(x_0)$$

Another way of interpreting the linear approximation is hence given by the following equation:

$$df \simeq \Delta(f)$$

that is, we approximate the variation $\Delta(f)$ of the function f through the variation df on the tangent line to the graph $y = f(x)$.

In the figure below, we see the two variations df on the left and $\Delta(f)$ on the right:

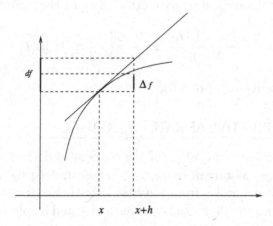

It is clear that, for small variations of x, that is, for small values of h, the two values df and $\Delta(f)$ are very close, while, if we consider large variations of h, the two values can be very different.

The linear approximation is very important when we want to estimate the *error* of a measurement. Let us see an example to clarify this important application.

Example 3.1.3 We compute the error we get for the volume V of a sphere, assuming a small measurement error of its radius r. Equivalently, we can also think to the same question, as the calculation of the change in the volume of a sphere, when we consider small changes in its radius.

Let us consider a sphere of radius $r_0 = 10$ cm with uncertainty of the measurement of 0.5%. This means that $\Delta r = (0.5/100) \times 10$ cm $= 0.05$ cm. From the formula of the linear approximation (3.2), we have that:

$$V(r) - V(r_0) \simeq V'(r_0)(r - r_0), \qquad \Delta(V) = V(r) - V(r_0), \qquad \Delta r = r - r_0$$

where the volume of the sphere is given by: $V(r) = 4\pi r^3/3$. We have immediately

$$V'(r) = 4\pi r^2$$

from which we can compute the volume change through the formula (3.2):

$$\Delta V = V(r) - V(r_0) \simeq 4\pi \cdot 10^2 \cdot 0.05 \simeq 63 \text{ cm}^3.$$

We can also estimate the relative error, that is, the error percentage:

$$\frac{V(r) - V(r_0)}{V(r)} \simeq \frac{63}{4\pi \; 10^3/3} \simeq 0.015.$$

We have a relative error of 1.5%.

3.2 THE DERIVATIVE AS RATE OF CHANGE

Another important application of the concept of derivative is through its interpretation as *rate of change*. To better understand this concept, we will look at examples from physics. Indeed, Newton developed the theory of limits and differential calculus motivated by physics problems. He was in fact interested in instantaneous velocity, that is, the rate of change of the distance as time varies and also to other related physics quantities such as acceleration, angular velocity and so on.

The average rate of change of a function $f : D \longrightarrow \mathbb{R}$ in an interval $(x_0, x_1) \subset D$ is given by the quotient of the variation of function in that interval and the length of the interval:

$$\frac{f(x_1) - f(x_0)}{(x_1 - x_0)}$$

We can also express it, more concisely, as

$$\text{rate of change} = \frac{\Delta f}{\Delta x}$$

Let us see a concrete example.

Example 3.2.1 Suppose a car is at the highway toll booth of "Roma Nord" at 8 am and then traveling on the highway arrives at the toll booth of "Firenze Sud" at 10:25 am. The two toll booths are 238 km apart. We want to determine the average rate of change of the distance with respect to the time. This will give us the average speed of the vehicle. We denote by $\Delta s = 238$ km the distance between the two toll booths. The car takes 2.25 hours to cover the distance Δs; hence, $\Delta t = 135/60 = 2.25$ hours. So the average speed is:

$$\frac{\Delta s}{\Delta t} = (238/2.25)\,\text{km/h} = 105.77\,\text{km/h}$$

Hence, the car traveled with an average speed of 105.77 km/h.

Let us now come back to the example of falling bodies that we have discussed in Chapter 1.

Example 3.2.2 The equation describing falling bodies is given by (see Chapter 1):

$$s(t) = s_0 + v_0 t - (1/2)g t^2, \qquad g = 9.8\,\text{m/s}^2$$

where $s(t)$ represents the position of the body at time t, while s_0 and v_0 represent the initial position and velocity, respectively. We now want to examine this equation more in detail.

The instantaneous velocity is the rate of change of the position, for infinitesimal intervals:

$$v(t) = \lim_{\Delta t \to 0} \frac{\Delta s}{\Delta t}$$

Hence, $v(t)$ represents the derivative of the position:

$$v(t) = \frac{ds}{dt} = v_0 - gt$$

We immediately see that, for $t = 0$, we have an initial velocity equal to v_0, justifying the statements in Chapter 1. Similarly, we can view the instantaneous acceleration $a(t)$ as the rate of speed variation with respect to time for infinitesimal time intervals. Thus, $a(t)$ is the derivative with respect to time of the instantaneous velocity:

$$a(t) = \frac{dv}{dt} = -g$$

Indeed, a falling body is subject to the acceleration of gravity, expressed here with the minus sign, as we have oriented the y-axis "upward".

On Earth $g = 9.8\,\text{m/s}^2$, while on other celestial bodies this value is different, but the law of falling bodies takes the same form.

3.3   LOCAL MAXIMA AND MINIMA

We now want to understand how the derivative of a function allows us to determine its local maxima and minima.

Definition 3.3.1 Let us consider a function $f : D \longrightarrow \mathbb{R}$, $x_0 \in D$. We say that f has *local maximum* $f(x_0)$ at x_0 if

$$f(x_0) \geq f(x), \quad \text{for all} \quad x \in I;$$

We say that f has *local minimum* $f(x_0)$ at x_0 if:

$$f(x_0) \leq f(x), \quad \text{for all} \quad x \in I$$

where I is a suitable open interval containing x_0, $I \subset D$.

Furthermore, if f is differentiable at x_0, we say x_0 is a *critical point* for f if:

$$f'(x_0) = 0$$

Geometrically, it is easy to see that, if $f(x_0)$ is a local maximum or minimum for the differentiable function f, then x_0 is a critical point. Indeed, we see that the tangent line at x_0 to the graph of f is horizontal. In other words, the slope of this line, given by the derivative of f at the point x_0, is zero.

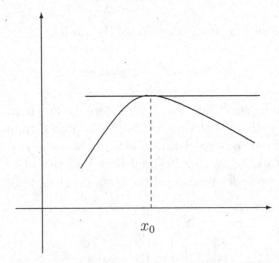

x_0

Theorem 3.3.2 Fermat's theorem on local maxima/minima: *Let $f : D \longrightarrow \mathbb{R}$ be a differentiable function. If $f(x_0)$ is a local minimum or a local maximum, then x_0 is a critical point of f.*

Proof. Suppose x_0 is a local maximum point, the case of local minimum is similar. Then, by definition, for each x in an interval I, we have $f(x_0) \geq f(x)$. We choose δ small enough to have: $(x_0 - \delta, x_0 + \delta) \subset I$ and we consider h such that $-\delta < h < +\delta$.

We have:

$$f(x_0) \geq f(x_0 + h), \qquad \text{that is} \qquad f(x_0 + h) - f(x_0) \leq 0$$

If $h > 0$, we have:

$$\frac{f(x_0 + h) - f(x_0)}{h} \leq 0.$$

If $h < 0$, we have:

$$\frac{f(x_0 + h) - f(x_0)}{h} \geq 0.$$

We compute the right-hand limit for h approaching zero (i.e. with $h > 0$):

$$f'(x_0) = \lim_{h \to 0^+} \frac{f(x_0 + h) - f(x_0)}{h} \leq \lim_{h \to 0^+} 0 = 0$$

We compute the left-hand limit for h approaching zero (i.e. with $h < 0$):

$$f'(x_0) = \lim_{h \to 0^-} \frac{f(x_0 + h) - f(x_0)}{h} \geq \lim_{h \to 0^-} 0 = 0$$

Since right-hand limit and left-hand limit coincide (f is differentiable), we have that $0 \leq f'(x_0) \leq 0$. Hence, $f'(x_0) = 0$. □

A word of warning: Fermat theorem does not state that all critical points are maxima or minima. In fact, as we see below, the derivative of f can be zero also at other points, which are neither maxima nor minima, called *inflection points* or *flexes*, for brevity.

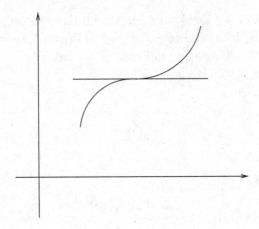

To understand if a critical point is a maximum, minimum or an inflection point, we need the following proposition.

Proposition 3.3.3 *Let* $f : D \longrightarrow \mathbb{R}$ *be a differentiable function and let* I *be an open interval in* D.

1. *If* $f'(x) > 0$ *for* $x \in I$, *then* f *is increasing in* I, *that is,* $f(x_2) > f(x_1)$ *for* $x_1 < x_2$, $x_1, x_2 \in I$.

2. *If* $f'(x) < 0$ *for* $x \in I$, *then* f *is decreasing in* I, *that is,* $f(x_2) < f(x_1)$ *for* $x_1 < x_2$, $x_1, x_2 \in I$.

The mathematical term for this property is f *strictly ascending or descending in* I, *but we prefer the shorthand terminology* f *"ascending" or "descending". When we take* $\geq$ *instead of* $>$ *(* $\leq$ *instead of* $<$ *),* f *is called monotone ascending or descending in* I. *We will not use this terminology.*

Proof. Suppose $f'(x) > 0$ for $x \in I$. By the Mean Value Theorem (see Appendix 3.8) given $x_1, x_2 \in I$, with $x_1 < x_2$, we have that, in the interval $[x_1, x_2]$, there exists a point c such that:

$$f'(c) = \frac{f(x_2) - f(x_1)}{x_2 - x_1}.$$

Hence:

$$f(x_2) - f(x_1) = f'(c)(x_2 - x_1) > 0 \quad \Longrightarrow \quad f(x_2) > f(x_1)$$

since $f'(c) > 0$ and $x_1 < x_2$, so that $x_2 - x_1 > 0$.

The case $f'(x) < 0$ is similar. $\qquad\square$

Let us see an example.

Example 3.3.4 We want to find local maximum and minimum points of the function: $f(x) = x^3 - 3x$. We compute the derivative of f: $f'(x) = 3x^2 - 3$ and we determine the *critical points*, that is, the points where the derivative is zero: $x = \pm 1$. To understand whether they are maximum, minimum or inflection points, we have to determine the sign of the derivative: $f'(x) \geq 0$ for $x \leq -1$, $x \geq 1$. So, we have that:

• in the intervals $(-\infty, -1)$ and $(1, \infty)$, the function is increasing;

• in the interval $(-1, 1)$, the function is decreasing.

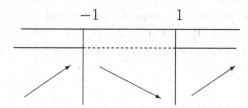

Hence, we have that $x = -1$ corresponds to a local maximum of f, while $x = 1$ corresponds to a local minimum of f. The values of f at these points are, respectively, $f(-1) = 2$ (maximum) and $f(1) = -2$ (minimum).

We conclude with the definition of global maximum and minimum of a function.

Definition 3.3.5 Let $f : D \longrightarrow \mathbb{R}$ be a function, $A \subseteq D$, x_0 in A. We say $f(x_0)$ is the *global maximum* of f in the set A, if $f(x) \leq f(x_0)$, for all $x \in A$. We say $f(x_0)$ is the *global minimum* of f in the set A, if $f(x) \geq f(x_0)$ for all $x \in A$.

We cannot always determine global maximum and minimum of f in a given $A \subseteq D$. For example, for $f : \mathbb{R} \longrightarrow \mathbb{R}$, $f(x) = x$, it is clear that f has no global maximum or minimum in the its domain.

But, even if we take a bounded interval A, we can encounter problems. For example, if we take the open interval $A = (0, 1)$ and the linear function $f(x) = x$, we see that there is no global maximum or minimum in A. However, this problem does not occur in the case of a continuous function and for A closed interval.

Theorem 3.3.6 Weierstrass theorem: *Every continuous function* $f : [a, b] \longrightarrow \mathbb{R}$ *has global maximum and minimum in* $[a, b]$.

This theorem states a very intuitive and geometrically clear fact; however, its proof requires sophisticated mathematical tools that are not available to us. We refer the interested reader to more specialized texts (see for example [3]).

Let us see an example of application.

Example 3.3.7 We want to find the global maximum and minimum points of the function: $f(x) = x^3 - 3x$ in the interval $[0, 3]$. In Example 3.3.4, we computed the local maximum and minimum values of f: ± 2 occurring in $x = \pm 1$. To determine the global maximum and minimum of f, we need to compute the value of f at the end-points of the interval, that is, at 0 and 3: $f(0) = 0$, $f(3) = 27 - 9 = 18$. Hence, the global minimum of f is $f(1) = -2$ and coincides with a local minimum. The global maximum is $f(3) = 18$.

3.4 GRAPH SKETCHING

We want to sketch the graph of a function, making use of all the information we gathered in our previous discussion regarding maxima, minima of a function and the role of the derivative.

Let us see the key steps to arrive effectively to sketch a graph by hand.

We explain each steps through a particular example:

$$f(x) = \frac{x^3}{x^2 - 4}$$

1. **Domain of f:** The first key information is to determine the domain of f, that is, the real values for which the function is defined. In the example of $f(x) = \frac{x^3}{x^2-4}$, the domain is given by $D = \mathbb{R} \setminus \{\pm 2\}$, that is, it consists of all real numbers, except the zeroes of the denominator, ± 2.

   Let us briefly recall that, for rational functions, the domain is given by the points for which the denominator is non-zero, while for algebraic functions containing even roots, the root argument must be positive. For logarithmic functions, we require the argument to be positive. Finally, polynomial, exponential and trigonometric functions are defined for all real numbers. We refer to Chapter 1 for a full treatment of the domain of the above-mentioned functions.

It is often useful to establish whether the function has *symmetries*. For example, we say that f is *even* if $f(x) = f(-x)$ or *odd* if $f(-x) = -f(x)$. This will allow us to sketch only part of the graph, recovering the rest by symmetry. In our special case since $f(-x) = -f(x)$, we will see this type of symmetry in action.

2. **Sign of** f: We determine the values of x for which $f(x) \geq 0$. In the example of $f(x) = \frac{x^3}{x^2-4}$, we have to determine the sign of the numerator and the denominator. The numerator is positive for $x \geq 0$, while the denominator is positive for $x < -2$ and $x > 2$. Hence, we have $f(x) \geq 0$ for $-2 < x \leq 0$, $x > 2$.

To determine the sign of a rational function, we need to solve two inequalities, one for the numerator and the other for the denominator. Then, we combine the results, as we did in the example. We recall that the exponential function is always positive, while to determine the sign of the logarithmic function, we need to determine for which values of x the argument of the logarithm is greater or equal to 1. For algebraic functions with odd roots, the sign is given by the sign of the function we are taking the root of, while even roots are always positive. For trigonometric functions, we use the unit circle to determine the sign, see Chapter 1 for more details.

We notice that we can also determine the intersections with the axes: the function intersects the x-axis and the y-axis only at the origin.

3. **Asymptotes of** f: We need to compute the limit of f for x approaching $\pm\infty$, in case the domain contains arbitrarily large values of x (positive or negative). Then, we need to take the limit at *boundary* points of the domain, (these are points that do not belong to the domain, but any open interval of such points intersects the domain of f). This allows us to get very important information about the graph of the function. Let us see the example of $f(x) = \frac{x^3}{x^2-4}$. The boundary points are $x = \pm 2$,

and it is necessary to compute the right-hand and left-hand limits separately, as the function changes sign at these points.

$$\lim_{x \to -2-} \frac{x^3}{x^2-4} = -\infty, \quad \lim_{x \to -2+} \frac{x^3}{x^2-4} = +\infty,$$

$$\lim_{x \to 2-} \frac{x^3}{x^2-4} = -\infty, \quad \lim_{x \to 2+} \frac{x^3}{x^2-4} = +\infty$$

When, as in this case, we have that the right/left-hand limit for x approaching a value $a \in \mathbb{R}$ is equal to $\pm\infty$, we say that the function has *vertical asymptote* $x = a$. So, the function $f(x) = \frac{x}{x^2-4}$ has vertical asymptotes $x = -2$ and $x = 2$.

We then compute the limits of f as x approaches $\pm\infty$:

$$\lim_{x \to -\infty} \frac{x^3}{x^2-4} = -\infty, \qquad \lim_{x \to +\infty} \frac{x^3}{x^2-4} = +\infty$$

For the calculation of these two limits, we used the theory regarding the order of infinity of Chapter 2.

We want to state the definitions of horizontal and oblique asymptotes and then see if the given function admits such asymptotes.

We say that the line $y = c$ is a *horizontal asymptote* for f if:

$$\lim_{x \to +\infty} f(x) = c \qquad \text{or} \qquad \lim_{x \to -\infty} f(x) = c$$

We call the line $y = mx + c$ an *oblique asymptote*, if the following three conditions are satisfied:

a. $\lim_{x \to +\infty} f(x) = \pm\infty \quad$ or $\quad \lim_{x \to -\infty} f(x) = \pm\infty$

b. $\lim_{x \to +\infty} \frac{f(x)}{x} = m \neq 0 \quad$ or $\quad \lim_{x \to -\infty} \frac{f(x)}{x} = m \neq 0$

c. $\lim_{x \to +\infty} [f(x) - mx] = c \quad$ or $\quad \lim_{x \to -\infty} [f(x) - mx] = c$

This happens, for example, when we have a rational function where the numerator has degree equal to the denominator degree plus one. But it can also happen in other cases with non-rational functions.

Let us see in our example:

$$\lim_{x\to+\infty} f(x)/x = \lim_{x\to+\infty} \frac{x^3}{(x^2-4)x} = 1, \qquad \lim_{x\to-\infty} f(x)/x = 1$$

Moreover:

$$\lim_{x\to+\infty} [f(x)-x] = \lim_{x\to+\infty} \frac{x^3}{(x^2-4)} - x = \lim_{x\to+\infty} \frac{x^3-x^3+4x}{(x^2-4)} = 0$$

and similarly

$$\lim_{x\to-\infty} [f(x)-x] = 0$$

Hence, we have that $y = x$ is the oblique asymptote.

4. **Local maxima and minima:** As we have seen in the previous section, to compute the maxima and minima of a function, it is necessary to compute the derivative and determine its sign.

$$f'(x) = \frac{x^2(x^2-12)}{(x^2-4)^2} \geq 0 \quad \Longrightarrow \quad x \leq -2\sqrt{3}, \quad x \geq 2\sqrt{3}$$

We know that, for the intervals where $f'(x) > 0$, the function is increasing, while for those where $f'(x) < 0$, the function is decreasing. The derivative is zero for $x = \pm 2\sqrt{3}$. To understand whether we have a maximum or a minimum, we have to determine the sign of the derivative: $f'(x) \geq 0$ for $x \leq -2\sqrt{3}$, $x \geq 2\sqrt{3}$. So, we have that:

- in the intervals $(-\infty, -2\sqrt{3})$ and $(2\sqrt{3}, \infty)$ the function is increasing;
- in the interval $(-2\sqrt{3}, 2\sqrt{3})$ the function is decreasing.

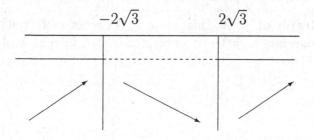

The point $x = -2\sqrt{3}$ is a local maximum point, while the point $x = 2\sqrt{3}$ is a local minimum point.

We also see that we have an *inflection point* for $x = 0$, i.e., a point at which the derivative is zero, but it is neither a maximum point nor a minimum point. So, in $x = 0$, the graph of f will have horizontal tangent line.

5. **Concavity:** The study of the sign of the second derivative $f''(x)$, i.e., the derivative of the function $f'(x)$, allows us to study the concavity of the function f. Indeed, if $f''(x) > 0$, the function is *convex* (positive concavity), if $f''(x) < 0$, the function is *concave* (negative concavity). This allows us to have important information on how to draw its graph. In our example:

$$f''(x) = \frac{8x(x^2 + 12)}{(x^2 - 4)^3} \geq 0 \quad \Longrightarrow \quad -2 < x \leq 0, x > 2$$

If $f''(x) > 0$, by the Proposition 3.3.3, the function $f'(x)$ is increasing, and hence, the tangent line to the graph has increasing slope, while, if $f''(x) < 0$, this slope will be decreasing. We summarize this behavior with the following diagram regarding the sign of the second derivative:

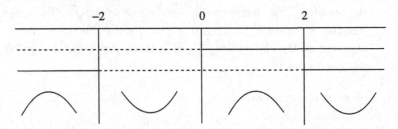

which is very useful, when we sketch the graph.

6. **Graph of f:** At this point, it is necessary to put together the information gathered so far and sketch the graph of f.

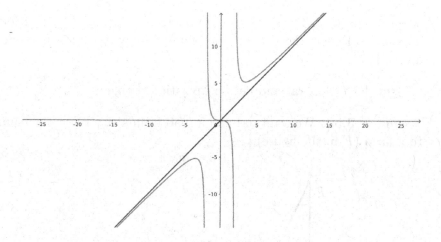

3.5   OPTIMIZATION

Optimization plays a key role in applied sciences and consists in finding the global maximum and minimum points of a function, called the *objective function* or alternatively *cost* or *loss function* in a domain determined by the concrete problem that we want to solve. Let us see an example.

Example 3.5.1 A shepherd has a 200 m of fencing wire and he wants to enclose a rectangular area as a pen for sheeps. What are the dimensions of the pen so that the fenced area is maximum?

Our objective function is represented by the fenced area, while we take as x and y the sides of the rectangle we want to fence. To proceed with the calculation of the maximum of the area function, we need to express it in terms of the variable x only. Since we are given the total length of the fencing wire, we can express y in terms of x (or vice-versa):

$$2(x+y) = 200 \implies y = 100 - x$$

Let us compute the objective function:

$$\text{Area} = xy = x(100 - x)$$

To determine the maximum, we need to compute maxima and minima of the objective function, in this case $A(x) = x(100 - x)$:

$$A'(x) = (100 - x) - x = 0 \implies x = 50 \implies y = 100 - 50 = 50$$

A quick check on the sign of $A'(x)$ shows us that $x = 50$ is a local maximum. Hence, the sides of the rectangle, giving us the maximum area, are $x = y = 50$.

Now let us look at another optimization problem.

Example 3.5.2 We want to build a road from a city (N in the figure) to a farm (F in the figure).

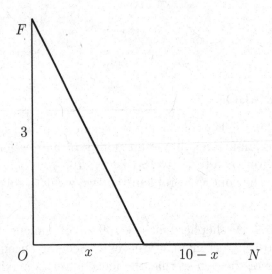

We already have a road at 3 km from the farm, represented by the line through the points O and N in the figure. Hence, we have that $OF = 3$ km. Also, we know that the distance between O and N is 10 km, that is, $ON = 10$ km. This road allows for a maximum speed of 5 km/h, while the road that we need to build, given the conformation of the territory, may be traveled only at a maximum speed of 4 km/h. We want to build the road so that we achieve the minimum travel time from the farm to the city. It is clear that building the road by looking at the shortest distance between the two points F and N will not lead us to minimize the travel time. The speed associated with the slow road is $v_1 = 4$ km/h, while the speed associated with the fast road is $v_2 = 5$ km/h. The goal of this optimization problem is to find the value of x (see figure) that minimizes the time necessary to get from the farm in F to the city in N. This is our objective function. Using the

Pythagorean theorem, we compute the length of the road with lower max speed:

$$d_1 = \sqrt{3^2 + x^2} = \sqrt{9 + x^2}$$

while the length of the other road is:

$$d_2 = 10 - x.$$

The time taken to travel each road is given by the quotient between the distance and the speed. Hence, the total time necessary to go from F to N is:

$$T(x) = \frac{d_1}{v_1} + \frac{d_2}{v_2} = \frac{\sqrt{9 + x^2}}{4} + \frac{10 - x}{5}.$$

In order to find the minimum for the objective function, we need to compute its derivative

$$T'(x) = \frac{x}{4\sqrt{9 + x^2}} - \frac{1}{5}$$

and set it equal to zero. We thus obtain the condition:

$$5x - 4\sqrt{9 + x^2} = 0 \implies 5x = 4\sqrt{9 + x^2}$$

Since we are dealing with positive quantities, such as distances, we can square without adding/losing solutions:

$$25x^2 = 16(9 + x^2)$$

from which

$$x^2 = \frac{144}{25 - 16}$$

hence

$$x = \pm\frac{12}{3} = \pm 4.$$

Let us now see which of these solutions is actually a minimum. We have to determine the sign of the derivative. We have:

$$T'(x) < 0 \quad \text{for } 0 < x < 4, \qquad T'(x) > 0 \quad \text{for } x > 4$$

Hence, $x = 4$ is a minimum of the objective function T. To minimize the travel time between the farm F and the town N, we have to build the road starting at 4 km from the point O.

Observation 3.5.3 In the optimization problems we examined so far, we see that the variable x belongs to a *closed interval*. This condition is extremely important. In fact, by Weierstrass Theorem 3.3.6, we have that a differentiable function defined in a closed interval always has global maximum and minimum in the interval. This is not true for a function defined in an *open interval*. For example, the function $f(x) = x$ does not have maximum or minimum in the interval $(0, 1)$, while evidently, it has a global maximum value 1 and a minimum value 0 in the closed interval $[0, 1]$. So, whenever we have an optimization problem for a function defined in an open interval, we cannot be certain that our problem admits a solution. We will not examine such cases in our present treatment, but it is important to know they may occur.

We conclude this section with an outline of the steps leading to the solution of an optimization problem.

- **Step 1**: Choose the problem variables. Often, we need a figure (the road to the farm, the shape of the fence, etc.), and it is important to understand what are the relevant quantities and how they depend from each other.

- **Step 2**: Determine the objective function. This is the function that we want to maximize or minimize (a length, an area, a time, etc.). It is important to express this function in terms of the variables chosen in step 1. But in the end, we need to use the relations in order to get a function of one variable only.

   In general, it is possible to address multivariate optimization problems, but we will not treat them here, because the techniques are essentially different.

- **Step 3**: Determine the global maxima and minima of the objective function through the calculation of the derivative and the evaluation at the end-points of the closed interval we are interested in (in case it is not obvious from the setup of the problem).

3.6 EXERCISES WITH SOLUTIONS

3.6.1 Use the linear approximation to estimate $(1.03)^{1/5}$.

Solution. We define the function $f(x) = x^{1/5}$. By the linear approximation formula:

$$f(1.03) \cong f(1) + 0.3f'(1) = 1 + 0.3 \times \frac{1}{5} = 1.006$$

Thus, $(1.03)^{1/5} \cong 1.006$.

3.6.2 Determine for which intervals the function $f(x) = \sin(x)\cos(x)$ is increasing.

Solution. The function f is the product of two trigonometric functions; hence, the domain is $\mathbb{R}$. Since f is periodic, we consider only $x \in [0, 2\pi]$.
   We determine the sign of $f'(x)$:

$$f'(x) = \cos^2(x) - \sin^2(x) = (\cos(x) + \sin(x))(\cos(x) - \sin(x)) \geq 0$$

We have:

$$\cos(x) \geq -\sin(x) \implies 0 \leq x \leq \frac{3\pi}{4}, \ \frac{7\pi}{4} \leq x \leq 2\pi$$

and

$$\cos(x) \geq \sin(x) \implies 0 \leq x \leq \frac{\pi}{4}, \ \frac{5\pi}{4} \leq x \leq 2\pi$$

Hence:

$$f'(x) \geq 0 \quad \text{when} \quad 0 \leq \frac{\pi}{4}, \ \frac{3\pi}{4} \leq x \leq \frac{5\pi}{4}, \ \frac{7\pi}{4} \leq x \leq 2\pi$$

These are the intervals for which f is increasing.

3.6.3 Determine the domain, the sign and the asymptotes of the function:

$$f(x) = x \log(2 - x^2)$$

Solution. The domain is given by the values of x such that $-\sqrt{2} < x < \sqrt{2}$. The sign is given by the solution of the inequalities $2 - x^2 > 1$, i.e., $-1 < x < 1$, and $x > 0$. So, the function is positive for $0 < x < 1$ and $x < -1$. We have two vertical asymptotes in $x = \pm\sqrt{2}$. There are no horizontal or oblique asymptotes.

3.6.4 Sketch the graph of the function $f(x) = \log(x^2 - 9)$.

Solution. We proceed step by step as we described in the Section 3.4.

1. **Domain:** The domain of f is given by the points where the argument of the logarithm is positive: $x < -3$, $x > 3$. The intersections with the x-axis are obtained by setting the argument of the logarithm equal to 1; hence, $x = \pm\sqrt{10}$. We note that the given function is even, that is: $f(x) = f(-x)$.

2. The sign of f is given by the values of x for which the argument of the logarithm is greater than 1. So, for $x \leq -\sqrt{10}$ and $x \geq \sqrt{10}$.

3. The points at the boundary of the domain of f are the values $x = \pm 3$ and $\pm\infty$. We hence need to compute the limit of f at each of these points. We immediately see that:

$$\lim_{x \to -3^-} \log(x^2 - 9) = -\infty, \qquad \lim_{x \to 3^+} \log(x^2 - 9) = -\infty,$$

$$\lim_{x \to -\infty} \log(x^2 - 9) = +\infty, \qquad \lim_{x \to +\infty} \log(x^2 - 9) = +\infty.$$

4. **Maxima and minima:** We have to compute the critical points, that is, the points in which the derivative is zero.

$$f'(x) = \frac{2x}{x^2 - 9} = 0$$

We have only the point $x = 0$, which does not belong to the domain. Hence, there are no local maxima or minima points.

5. **Concavity:** For the concavity, we compute the second derivative:

$$f''(x) = -2\frac{x^2 + 9}{(x^2 - 9)^2}$$

Thus, there are no inflection points; the function has negative concavity at all points.

6. Graph:

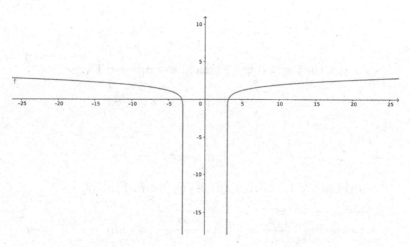

3.6.5 Sketch the graph of the function $f(x) = -xe^{-\frac{1}{x}}$.

Solution. We proceed step by step as we have described in Section 3.4.

1. **Domain:** $\forall x \neq 0$ i.e. $\mathbb{R} \setminus \{0\}$.

   As for the symmetries: f is neither an even function, because $f(-x) = xe^{\frac{1}{x}} \neq f(x)$, nor an odd function, because $f(-x) = xe^{\frac{1}{x}} \neq -f(x) = xe^{-\frac{1}{x}}$.

   There are neither intersections with the y-axis, since the point $x = 0$ is not in the domain of f, nor with the x axis, since $0 = -xe^{-\frac{1}{x}}$ implies that $x = 0$.

2. **Sign of f:** $f(x) > 0$ if $x < 0$ and $f(x) < 0$, if $x > 0$, since $e^{-\frac{1}{x}} > 0$.

3. **Asymptotes:**
$$\lim_{x \to 0^+} -xe^{-\frac{1}{x}} = 0$$
$$\lim_{x \to 0^-} -xe^{-\frac{1}{x}} = 0 \cdot \infty$$
   which is an indeterminate form.
   By making the change of variable $\frac{1}{x} = t$, we get $1 = tx$ and $x = \frac{1}{t}$ and then:
$$\lim_{x \to 0^-} -xe^{-\frac{1}{x}} = \lim_{t \to -\infty} -\frac{1}{t}e^{-t} = -\frac{+\infty}{-\infty}$$

which is still an indeterminate form. By De l'Hospital's Rule:

$$\lim_{t \to -\infty} \frac{-(e^{-t}) \cdot (-1)}{1} = \lim_{t \to -\infty} e^{-t} = +\infty = \lim_{x \to 0^-} -xe^{-\frac{1}{x}}$$

We get that $x = 0$ is a vertical asymptote. Then

$$\lim_{x \to +\infty} -xe^{-\frac{1}{x}} = -\infty$$

and

$$\lim_{x \to -\infty} -xe^{-\frac{1}{x}} = +\infty$$

so there are no horizontal asymptotes. Finally,

$$m = \lim_{x \to \pm\infty} \frac{f(x)}{x} = \lim_{x \to \pm\infty} \frac{-xe^{-\frac{1}{x}}}{x} = \lim_{x \to \pm\infty} -e^{-\frac{1}{x}} = -e^0 = -1$$

$$c = \lim_{x \to \pm\infty} -xe^{-\frac{1}{x}} + x = -\infty + \infty$$

which is an indeterminate form.
Similarly :

$$c = \lim_{x \to \pm\infty} x(-e^{-\frac{1}{x}} + 1) = \pm\infty \cdot 0.$$

Doing the substitution $\frac{1}{x} = t$ as before, we find:

$$c = \lim_{t \to 0^\pm} \frac{1 - e^{-t}}{t}$$

which, by De l'Hospital's rule:

$$c = \lim_{t \to 0^\pm} \frac{e^{-t}}{1} = 1$$

So $y = -x + 1$ is an oblique asymptote.

4. **Maxima and minima:** We compute the derivative:

$$f'(x) = -e^{-\frac{1}{x}} \left(\frac{x+1}{x} \right)$$

The function f is increasing for $-1 < x < 0$ and decreasing for $x < -1$, $x > 0$ ($e^{-\frac{1}{x}} > 0$). Moreover, the point $x_0 = -1$ is a minimum point with $f(-1) = e$.

5. **Concavity:** Let us compute the second derivative:

$$f''(x) = -\frac{e^{-\frac{1}{x}}}{x^3}$$

We have that $f''(x) > 0$ if $x < 0$ and $f''(x) < 0$ if $x > 0$ ($e^{-\frac{1}{x}} > 0$). The point with $x = 0$ is not an inflection point, since it does not belong to the domain of f.

Graph.

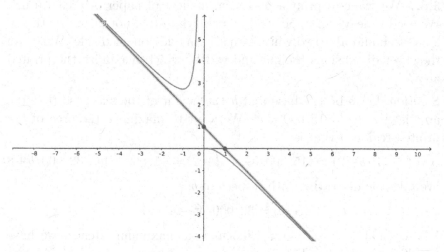

3.6.6 A landlord is managing an apartment building with 50 units. The monthly rent for an apartment is 800 euros. For every 50 euro increase of the monthly rent, we estimate that an apartment becomes available, since the tenants will try to get a cheaper accommodation. For example, an increase of 150 euro/month, will result in three vacancies. We have 100 euros per month of fixed expenses, like gardening and water. We want to determine the monthly rent that maximizes the landlord profit.

Solution. We need to express the objective function, in this case the profit P, as a function rent a and other variables as monthly expenses. We also need to take into account that, every 50 euro increase of the rent, we get one vacancy.

Hence, the profit P, as a function of the monthly rent a, is given by:

$$P(a) = (a - 100)\left(50 - \frac{a - 800}{50}\right).$$

To find the maximum of the function $P(a)$, we need to compute the derivative:

$$P'(a) = 50 - \frac{a - 800}{50} - \frac{a - 100}{50} = 0$$

We then obtain

$$3400 - 2a = 0$$

giving $a = 1700$. By looking at the positivity of the derivative, we can see that this is a maximum for the objective function.

3.6.7 We want to print a poster on a sheet of paper of area 1.5 m². We need to leave an upper and a lower white (i.e. not printed) strip of 2 cm each and also two white strips 1 cm wide on each side. What are the sheet dimensions (height and width) which maximize the printed area?

Solution. Let b be the base and h the height of the sheet, so that the area is $A = bh = 15,000$ cm². We want to maximize the area of the printed region S that is:

$$S = (h - 4)(b - 2) = (15,000/b - 4)(b - 2) = 15,000 - 4b - 30,000/b + 8$$

We take the derivative with respect to b:

$$S'(b) = 30,000/b^2 - 4 = 0$$

We see that $b = \sqrt{7500} = 86.5$cm is a maximum. Hence, we have $b = 86.5$ cm and $h = 173$ cm.

3.7 SUGGESTED EXERCISES

3.7.1 Use the linear approximation to approximate the following numbers without the use of a calculator.

1. $\sqrt{0.98}$

2. $0.97 \log(0.97)$

3. $e^{-0.02}$

4. $(1.03)^{1/4}$.

3.7.2 A rock falls from a hill top and hits the ground after 7.5 seconds. Determine the hill height and the speed when the rock hits the ground.

3.7.3 A brick detaches from a building and after 0.98 second hits the head of a person 1.75 m high. Determine the height of the building and the speed of the brick al moment of the impact.

3.7.4 Determine the area error if we allow an error of 0.01 in the measurement of the side ℓ of the following geometric objects:

1. A square with side ℓ.

2. A rectangle with sides ℓ and 2ℓ.

3. An equilateral triangle with side ℓ.

3.7.5 A rectangle is in the first quadrant of the Cartesian plane with two of its sides on the coordinate axes and one vertex at the origin. The vertex opposite to the origin is on the curve $y = 1/x$. Find the sides of the rectangle that minimize its perimeter.

3.7.6 Assume we have an error of 0.04 in measuring the radius of a semicircular window of radius 50 cm. Compute, using the linear approximation, the percentage error of the area measurement.

3.7.7 A square-shaped window is installed surmounted by a semicircle one. The base of the window is 60 cm; however, the measurement has a possible 0.3 cm error. Determine the maximum error in calculation of the windows area, using the linear approximation.

3.7.8 1) Determine the domain and the asymptotes of the following functions.

1. $f(x) = \frac{x}{e^{x^2} - 3}$.

2. $f(x) = \frac{e^{x^2 - 1}}{x + 1}$.

3. $f(x) = \frac{\log(x^2 + 2x)}{x - 1}$.

4. $f(x) = \log(x^2 - 8x)$

5. $f(x) = \sqrt{\frac{x - 3}{x^2 - 1}}$

6. $f(x) = \frac{\sin(x)}{\cos(x)-1}$

7. $f(x) = (1 - 3x^2)^{1/4} - e^{-2x}$

8. $f(x) = \frac{\sqrt{1-x}}{\log(x^2+1)}$

9. $f(x) = \frac{2x-3x^3-5}{3x^2-x^3+2x}$

10. $f(x) = \frac{3x(x-2)}{x^2-4}$

11. $f(x) = \frac{x^4+x}{1+x^2-2x}$

12. $f(x) = \frac{\log(x+3)}{x^2-1}$

13. $f(x) = \sqrt{7x - 2}$

14. $f(x) = \frac{e^{2x^2}-1}{x^2}$

15. $f(x) = \frac{2x(x-4)}{x^2-9}$

3.7.9 For each of the following functions, specify the intervals where the function is continuous and/or differentiable.

1. $f(x) = x^3 - 3x^2 + 6$

2. $f(x) = x^4 - 2x^2 + 8$

3. $f(x) = x\sin(x)$

4. $f(x) = \frac{x-3}{x^2-4}$

5. $f(x) = 3e^{-2x}$

6. $f(x) = \log(x^2 - 9)$

7. $f(x) = \log^2(x)$

8. $f(x) = (1 - |x|)^2$

9. $f(x) = 4\log\left(\frac{1}{2x}\right)$

10. $f(x) = e^{\sqrt{x}} + \sqrt{2}$

3.7.10 For each of the given functions determine the intervals of the domain where the function is increasing.

1. $f(x) = x^3 - 3x^2 + 6$

2. $f(x) = 3e^{-2x}$

3. $f(x) = \log(x^2 - 9)$

4. $f(x) = \log^2(x)$

3.7.11 Sketch the graph of the following functions:

1.
$$y = \frac{x^2 - 4}{x + 1}$$

2.
$$y = \frac{x^2 - 2}{x^2 + 4}$$

3.
$$y = \log \frac{x + 1}{x - 1}$$

4.
$$y = \log(x^2 - 2x)$$

5.
$$y = e^{\frac{2x^2 - 1}{x - 1}}$$

6.
$$y = \frac{e^{x-2}}{x + 3}$$

7.
$$y = \frac{\sin x}{\cos x - 1}$$

8.
$$y = \frac{x - 1}{x + 1}$$

3.7.12 We want to build three identical rectangular cellars. Building a meter of wall costs 500 euros. We have a total of 24,000 euros available and we do not take into account the cost of the roof. What are the dimensions in order to have the maximum volume of the cellars?

3.7.13 Prove that the square is the rhomboid of maximum area with given perimeter.

3.7.14 Find two numbers whose sum is 20 and for which the sum of squares is minimal.

3.7.15 Let $R = 30$ cm be the radius of a sphere. Find the radius r and height h of the cylinder inscribed in the sphere of maximum lateral area.

3.7.16 Given two positive numbers whose product is k, prove that the sum of their cubes is minimal when the two numbers are equal.

3.7.17 A lifeguard, at the beach, sees a swimmer in the water in need of help. The swimmer is 20 m away from the shore, and, along the beach, 60 m away from the lifeguard. The lifeguard runs at 10 km/h and swims at 5 km/h. At what point should she jump into the water to reach the swimmer as quickly as possible?

3.7.18 Suppose we want to build an aluminum box of volume 125 cm³ of height h, with a square base of side b. What are the dimensions of b and h that minimize the *total area* of the box?

3.7.19 Suppose we have a fencing wire 12 m long and we want to build two small pens, one square shaped, the other round shaped. What are the perimeters of the two fences, so that the fenced area is *minimum*?

3.7.20 In a company the costs for monthly production consists of fixed costs of 1000 euro and variable costs according to the amount of goods q, which the company produces every month. The monthly variable costs follow the law

$$C(q) = 12q^2 - 960q.$$

The monthly revenue depends on q and it is given by $R(q) = 10q^2$. Find q to get the maximum profit.

3.7.21 We want to build a column consisting of a cylinder surmounted by a hemisphere having its base on the cylinder. What are the dimensions of the column knowing that its lateral area is 147π dm² and we want the largest possible volume?

3.8 APPENDIX: THEOREMS OF DIFFERENTIAL CALCULUS

In this appendix we give the statements and the proof of two fundamental theorems of the differential calculus: Rolle's Theorem and Lagrange Theorem. They are the key to our methods to determine the local maxima and minima of a function.

We start with Rolle's theorem.

Theorem 3.8.1 Rolle's theorem: *Let $f : [a, b] \to \mathbb{R}$ be a continuous function in $[a, b]$, which is differentiable in (a, b). Assume $f(a) = f(b)$. Then, there exists a point $c \in (a, b)$ with $f'(c) = 0$.*

Proof. Since f is continuous, f admits a global maximum M and minimum m in $[a, b]$ (by Weierstrass Theorem 3.3.6). Hence we have that: either the maximum and the minimum are reached in one of the end-points of the closed interval $[a, b]$, or one of them is reached at a point belonging to the open interval (a, b). We consider the first case only, since in the second case, we apply Fermat Theorem to obtain $c \in (a, b)$, such that $f'(c) = 0$.

If maximum and minimum are both reached at the end-points of the interval $[a, b]$, since by our hypothesis $f(a) = f(b)$, we immediately obtain $M = m$. So, f is constant in $[a, b]$ and hence its derivative is zero at every point c in $[a, b]$. □

By Rolle Theorem, we can prove the *Mean Value Theorem* also called the *Lagrange Theorem*. A crucial consequence of this theorem is the following. When the derivative of a function is positive for all x in an interval, then the function is increasing in the interval (see Proposition 3.3.3).

Theorem 3.8.2 Lagrange theorem: *Let $f : [a, b] \to \mathbb{R}$ be a continuous function on the closed interval $[a, b]$ and differentiable in the open interval (a, b). Then, there exists $c \in (a, b)$ such that*

$$f'(c) = \frac{f(b) - f(a)}{b - a}.$$

Proof. Define g and h as follows:

$$g(x) = f(a) + \frac{f(b) - f(a)}{b - a}(x - a), \qquad h(x) = f(x) - g(x)$$

Both g and h are continuous in $[a, b]$ and differentiable in (a, b). We immediately see that

$$h(a) = f(a) - g(a) = 0 \qquad h(b) = f(b) - g(b) = 0$$

So let us apply Rolle's theorem to the function h. So, there exists $c \in (a, b)$ with $h'(c) = 0$, that is $f'(c) = g'(c)$. Hence:

$$f'(c) = g'(c) = \frac{f(b) - f(a)}{b - a}$$

□

CHAPTER 4

Integrals

4.1 THE DEFINITE INTEGRAL

The definite integral answers to the very natural question regarding the computation of the area between the graph of a positive function $f : [a, b] \longrightarrow \mathbb{R}$ and the x-axis.

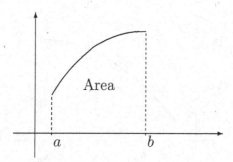

Let us now approximate such area, through the concept of *Riemann sums*. We answer to this question, first intuitively, through some geometrical reasoning, and then, we formulate the rigorous mathematical notions.

Suppose we have an increasing function f, that is, $f(x_1) < f(x_2)$, $x_1 < x_2$. Assume furtherly that f is positive (i.e. $f(x) > 0$) and continuous in the interval $[a, b]$. We divide this interval in n equal parts, that is, intervals of length $\Delta x = (b - a)/n$. We construct the sum s_n of the rectangles, shown in the picture below,

$$s_n = f(x_0)\Delta x + f(x_1)\Delta x + \cdots + f(x_{n-1})\Delta x$$

DOI: 10.1201/9781003343288-4

Since the function f is increasing, the sum s_n corresponds to the sum of the areas of the rectangles between the graph of the function f and the x-axis in the closed interval $[a, b]$.

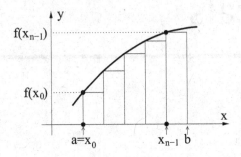

Similarly, we can construct the sum S_n obtained by taking the rectangles with height the right endpoint of each interval, instead of the left one as we did for s_n:

$$S_n = f(x_1)\Delta x + f(x_2)\Delta x + \cdots + f(x_n)\Delta x$$

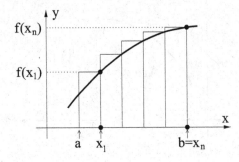

Geometrically, we see from the graph that:

$$s_n \leq \text{Area between } y = f(x) \text{ and the } x-\text{axis} \leq S_n$$

We also see that, when n becomes large, the value of s_n gets closer to the value of S_n and both approximate the area between $y = f(x)$ and the x-axis. We also note that, for very large n, it is not important whether we take the value of f in the left or right endpoint of each interval $[x_i, x_{i+1}]$: the sum of the areas of the rectangles will in any case approximate the desired area.

This leads us to give the following definition of definite integral.

Definition 4.1.1 Let $f : [a, b] \longrightarrow \mathbb{R}$ be a bounded function, i.e., $|f(x)| < M$ for every x in $[a, b]$. Let us divide $[a, b]$ into intervals $[x_i, x_{i+1}]$, each of length Δ_i, and choose a value t_i in each interval $[x_{i-1}, x_i]$:

$$a = x_0 \leq t_1 \leq x_1 \leq t_2 \leq x_2 \leq \ldots \leq x_{n-1} \leq t_n \leq x_n = b.$$

We define *Riemann sum*:

$$R_n = f(t_1)\Delta_1 + f(t_2)\Delta_2 + \cdots + f(t_n)\Delta_n$$

We say that f is *integrable* in the interval $[a, b]$, if the limit:

$$\lim_{\Delta \to 0} R_n \tag{4.1}$$

exists and it is a real number (i.e. the limit is not $\pm\infty$), where Δ is the maximum of the lengths of Δ_i as i varies.

We denote the value of this limit as:

$$\int_a^b f(x)dx := \lim_{\Delta \to 0} R_n$$

and we call it *the definite integral of f in the interval $[a, b]$* or shortly *the integral of f from 0 to 1*.

Sometimes, when it is clear from the context, we will omit the word "definite".

The sum R_n is called *Riemann sum*.

We observe in the previous example that, according to our definition, we have immediately:

$$s_n \leq R_n \leq S_n$$

Thus, the definite integral of a positive and increasing continuous function in an interval $[a, b]$ represents geometrically the area between the graph of the function and the x-axis. In general, for a positive function, not necessarily increasing, we can give the same geometric interpretation: the definite integral represents the area between the given function and the x-axis. When the function is negative, to find the area between its graph and the x-axis, it is enough to change the sign and compute the definite integral. However, for a function, which is not continuous, we may lose this geometric interpretation. In the next observation, we take into exam an example of this situation. These reasonings should be taken as an opportunity to think deeper into concepts, regarding integration, that we are unable to fully explore in our discussion.

Observation 4.1.2 The definition of definite integral is given through the concept of limit; however, we know that, in general, a limit may not exist. Indeed, we are going to see an example of a bounded function in a closed interval for which the $\lim_{\Delta \to 0} R_n$ does not exist, and hence, it is not possible to compute its definite integral in such interval.

Consider the *Dirichlet function* $f : [0,1] \longrightarrow \mathbb{R}$:

$$f(x) = \begin{cases} 1 & x \quad \text{rational} \\ 0 & x \quad \text{irrational} \end{cases}$$

Depending on how we choose the points $t_1, \ldots, t_n$ in the intervals $[x_i, x_{i+1}] \subset [0,1]$, we have that:

- $R_n = 1$, if we choose all t_i rational,

- $R_n = 0$, if we choose all t_i irrational.

It is then clear that the limit (4.1) does not exist. So, the given function, though bounded in the interval $[0,1]$, is not integrable in this interval.

We have an important result, whose proof is in Appendix 4.12.

Proposition 4.1.3 *A continuous function is integrable in a closed interval.*

Hence, all the most interesting functions in applied sciences, such as polynomial, rational, algebraic, exponential and trigonometric functions that we studied in Chapter 1, are integrable in every closed interval in their domain.

At this point, however, we should not assume that continuity is a necessary condition to have integrability; we will see an example in the next section (Example 4.2.3).

The definition of integral is not very helpful for actual computations; let us see an example on how we can apply directly the definition to compute a very simple integral. We will see, later on, much more effective methods for computing integrals.

Example 4.1.4 We want to compute the integral of $f(x) = x$ from 0 to 1. From the elementary geometry, we know that the area between the line $y = x$ and the x-axis, for x in the interval $[0,1]$, is the area of a right triangle with two sides equal to 1. So, our geometric and intuitive reasoning shows that the definite integral of $f(x) = x$ is $1/2$. Let us proceed now to the calculation through the definition. We take $\Delta_i = (1-0)/n = 1/n$ as the width of the intervals and calculate R_n taking the point x_i as the left end-point of the ith interval. Hence:

$$R_n = f(x_0)\Delta x + f(x_1)\Delta x + \cdots + f(x_{n-1})\Delta x =$$

$$= 0 \cdot \frac{1}{n} + \frac{1}{n} \cdot \frac{1}{n} + \frac{2}{n} \cdot \frac{1}{n} + \cdots + \frac{n-1}{n} \cdot \frac{1}{n} =$$

$$= \frac{1+2+\ldots n-1}{n^2} = \frac{n(n-1)}{2n^2}$$

where we have used the formula for the sum of the first n natural numbers. Now, let us compute the limit as n tends to infinity of the function $\frac{n(n-1)}{2n^2}$:

$$\lim_{n \to +\infty} \frac{n(n-1)}{2n^2} = \frac{1}{2}$$

using the methods studied in Chapter 2.

We have:

$$\int_0^1 f(x)dx = \int_0^1 x\,dx = 1/2$$

and we see immediately that this value coincides with the area of the right triangle with two sides equal to 1. However, this method is not very efficient for the calculation of areas or the definite integral of more complicated functions.

In Section 4.3, we will see an effective method for computing the integral that does not make use of the definition.

4.2   PROPERTIES OF THE DEFINITE INTEGRAL

In this section, we give some properties of the definite integral, which are obtained, almost immediately, from the corresponding properties of limits.

Proposition 4.2.1 *Let f and g be two integrable functions in the interval $[a, b]$. Then, $f + g$ and kf, where k is constant, are integrable functions in the same interval, and we have:*

$$\int_a^b [f(x) + g(x)]\, dx = \int_a^b f(x)\, dx + \int_a^b g(x)\, dx$$

$$\int_a^b k f(x)\, dx = k \int_a^b f(x)\, dx.$$

Proof. Let us see the first of the two equalities, leaving the second one as an exercise.

$$\int_a^b [f(x) + g(x)]\, dx = \lim_{\Delta \to 0} R_n =$$

$$= \lim_{\Delta \to 0} \{[f(t_1) + g(t_1)]\Delta_1 + \cdots + [f(t_n) + g(t_n)]\Delta_n\} =$$

$$= \lim_{\Delta \to 0} [f(t_1)\Delta_1 + \cdots + f(t_n)\Delta_n + g(t_1)\Delta_1 + \cdots + g(t_n)\Delta_n] =$$

$$= \lim_{\Delta \to 0} [f(t_1)\Delta_1 + \cdots + f(t_n)\Delta_n] +$$

$$+ \lim_{\Delta \to 0} [g(t_1)\Delta_1 + \cdots + g(t_n)\Delta_n] =$$

$$= \int_a^b f(x)dx + \int_a^b g(x)dx$$

since the limit of the sum is the sum of the limits. $\square$

Let us look at another proposition, which will prove to be very useful for the exercises. The proof is a simple application of the concepts seen so far, and we leave it to the reader.

Proposition 4.2.2 *Let us consider an integrable function f in the interval $[a, c]$ and let b such that $a < b < c$. We have:*

$$\int_a^c f(x)\, dx = \int_a^b f(x)\, dx + \int_b^c f(x)\, dx$$

Now we can give an example of a bounded, integrable function, which is *not* continuous, as we mentioned in the previous section, when discussing the Dirichlet function.

Example 4.2.3 Let us consider the function $f : [0, 1] \longrightarrow \mathbb{R}$:

$$f(x) = \begin{cases} 1 & x < \frac{1}{2} \\ 0 & x \geq \frac{1}{2} \end{cases}$$

We see that it is an integrable function; in fact, according to the previous proposition, we have:

$$\int_0^1 f(x)\,dx = \int_0^{1/2} f(x)\,dx + \int_{1/2}^1 f(x)\,dx = 1/2$$

where we have used the fact that the integral of a constant function $f(x) = k$ in an interval $[a, b]$ is given by:

$$\int_a^b k\,dx = k(b - a)$$

that is, the area of the rectangle as in the figure below.

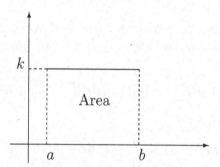

We leave this easy check to the reader as an exercise.

For a function f defined in an interval $[a, b]$, we have the following statements:

$$f \text{ continuous} \implies f \text{ integrable and } f \text{ bounded}$$

$$f \text{ integrable} \;\not\!\!\implies\; f \text{ continuous}$$

$$f \text{ bounded} \;\not\!\!\implies\; f \text{ integrable}$$

We invite the student to think about each implication keeping in mind the above examples.

We conclude this section with a proposition, which will turn out to be very useful in applications.

Proposition 4.2.4 *Let f and g be two integrable functions defined in an interval $[a, b]$ and such that $f(x) \leq g(x)$. Then:*

$$\int_a^b f(x)\, dx \leq \int_a^b g(x)\, dx$$

Proof. We apply the definition of integral and look at the sums R_n for each of the two functions:

$$f(t_1)\Delta_1 + \cdots + f(t_n)\Delta_n \leq g(t_1)\Delta_1 + \cdots + g(t_n)\Delta_n$$

Taking the limit:

$$\int_a^b f(x)\, dx = \lim_{\Delta \to 0} f(t_1)\Delta_1 + \cdots + f(t_n)\Delta_n \leq$$

$$\leq \lim_{\Delta \to 0} g(t_1)\Delta_1 + \cdots + g(t_n)\Delta_n = \int_a^b g(x)\, dx$$

$\square$

From this result, together with the observation regarding the integral of a constant function, see Example 4.2.3, we immediately have the following corollary.

Corollary 4.2.5 *Let f be an integrable function defined on an interval $[a, b]$ such that $m \leq f(x) \leq M$. Then:*

$$m(b - a) \leq \int_a^b f(x)\, dx \leq M(b - a)$$

This corollary expresses the fact that the integral of a function f in a interval $[a, b]$ has a value between the values of the two areas indicated in the figure:

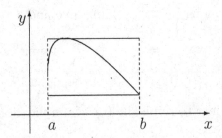

If the function f is continuous, m and M can be chosen as the maximum and minimum of f, respectively, in the given interval.

4.3 THE FUNDAMENTAL THEOREM OF CALCULUS

The fundamental theorem of calculus, also known as Torricelli's theorem, allows us to connect concepts that are apparently very different from each other: the calculation of the area between the graph of a function and the x-axis and the antiderivative of the function. Let us start with a definition.

Definition 4.3.1 Let $f: [a, b] \to \mathbb{R}$ be a continuous function. We say that a function $F : [a, b] \longrightarrow \mathbb{R}$ is the *primitive* or *antiderivative* of f if:

$$F'(x) = f(x) \qquad \text{for all} \quad x \in [a, b].$$

It is important to note that, given a function f and its primitive F, we have that $F + c$ is also a primitive of f, for any constant c. Indeed, $D(F(x) + c) = D(F(x)) = f$. Hence, the primitive of a given function is not unique. We denote with $\int f(x)dx$ the "generic" primitive of f. For example, if $f(x) = x$, we write:

$$\int x dx = \frac{x^2}{2} + c$$

This notation expresses the fact that every function $F(x) = \frac{x^2}{2} + c$ is a primitive of x for a given constant c. We call $\int f(x)dx$ the *indefinite integral* of f, which is called the *integrand* function.

Let us now look at the most important result of this chapter, whose proof can be found in Appendix 4.12.

Theorem 4.3.2 Fundamental theorem of calculus:
Let $f: [a, b] \to \mathbb{R}$ be a continuous function with primitive F. Then,

$$\int_a^b f(x)dx = F(b) - F(a).$$

Let us now see how the previous theorem allows us to compute the definite integrals more easily, with respect to our calculations in the previous section.

Example 4.3.3 We compute the integral of $f(x) = x$ in the interval $[0, 1]$, using the Fundamental Theorem of Calculus. We know that a primitive of f is given by $F(x) = x^2/2$; therefore:

$$\int_0^1 f(x)dx = F(1) - F(0) = 1/2$$

This result agrees with what we obtained in Example 4.1.4, but with more effort.

Notation: When we compute a definite integral, it is customary to denote the difference $F(b) - F(a)$ more concisely as $[F(x)]_a^b$.

The following proposition allows us to compute the integral of a sum of functions, and it is an immediate consequence of the derivation properties.

Proposition 4.3.4 *Let $f, g : [a, b] \longrightarrow \mathbb{R}$ be continuous functions. Then:*

1. $\int [f(x) + g(x)]dx = \int f(x)dx + \int g(x)dx$

2. $\int kf(x)dx = k \int f(x)dx$ where k is a constant.

Let us see how this proposition allows us to compute the integral of polynomials or other simple functions.

Before proceeding, we write the primitives of some functions, which we leave to the reader as a straightforward check:

- $\int x^n dx = \frac{x^{n+1}}{n+1} + c, \, n \neq -1$

- $\int e^x dx = e^x + c$

- $\int \sin x dx = -\cos x + c$

- $\int \cos x dx = \sin x + c$

In Appendix 4.11, we give a more exhaustive list of the primitives of the most important functions and some integration methods, which we will see later on in this chapter.

Example 4.3.5 We want to compute the integral:

$$\int_0^1 (2x^5 - 7x^3 + 2x)dx = 2\int_0^1 x^5 dx - 7\int_0^1 x^3 dx + 2\int_0^1 xdx$$

Notice that we have used the two properties stated in the previous proposition regarding the sum of functions and the product of a function by a constant. We proceed applying Torricelli's theorem:

$$\int_0^1 (2x^5 - 7x^3 + 2x)dx =$$

$$= 2\left[\frac{x^6}{6}\right]_0^1 - 7\left[x^4/4\right]_0^1 + 2\left[x^2/2\right]_0^1 = 2/6 - 7/4 + 1 = -5/12$$

Now we look at an example involving a trigonometric function.

Example 4.3.6 We want to compute:

$$\int_0^\pi \sin(x)dx$$

Applying Torricelli's theorem, we see that:

$$\int_0^\pi \sin(x)dx = [-\cos(x)]_0^\pi = 2$$

Now let us see what happens, if we consider the integral in the interval $[0, 2\pi]$:

$$\int_0^{2\pi} \sin(x)dx = [-\cos(x)]_0^{2\pi} = 0$$

This appears as a contradiction, because certainly we have a non-zero area between the graph function $\sin(x)$ and the x-axis. However, taking a closer look at this integral, we notice that the function $\sin(x)$ is positive for x between 0 and π, while it is negative for x between π and 2π. Since we are adding up equal quantities, but with opposite sign, the integral is zero. Therefore, when we want to actually compute an area and not simply a definite integral, we must always make sure that the function that we want to integrate, is positive; thus, we need to multiply it by a sign whenever it is negative. This is equivalent to take the function in absolute value:

$$\text{Area} = \int_0^{2\pi} |\sin(x)|dx = [-\cos(x)]_0^\pi - [-\cos(x)]_\pi^{2\pi} = 4$$

4.4   INTEGRATION BY SUBSTITUTION

The Fundamental Theorem of Calculus allows us to compute the definite integral of a function through the calculation of its primitive, that is, via an *indefinite integral*. In this and the next sections, we turn to the description of effective methods for the calculation of indefinite integrals.

Let us recall the chain rule, which is the basic idea we will use to do integration by substitution:

$$D[f(g(x))] = f'(g(x))g'(x) \qquad (4.2)$$

For example, if we consider the function $f(x) = e^{3x}$, the chain rule gives us:

$$D[e^{3x}] = 3e^{3x}$$

We now want to integrate $f(x) = e^{3x}$ by means of the chain rule as in (4.2), in order to obtain the primitive as a composition of functions:

$$\int e^{3x}\,dx = (1/3)\int (3e^{3x})\,dx = (1/3)e^{3x} + c$$

In general, to solve an integral by substitution, we need to go through the following steps:

- **Step 1:** We choose the function $g(x)$ (see formula (4.2)).

- **Step 2:** We set $u = g(x)$ and $du = g'(x)dx$.

- **Step 3:** We *substitute* u and du in the indefinite integral. If we get a integral with just u, then we can proceed to step 4. Otherwise, we have to stop, as the chosen substitution is not suitable.

- **Step 4:** We solve the indefinite integral in the new variable u.

- **Step 5:** We *substitute* again $u = g(x)$, so as to obtain a result depending on x only.

Let us see the procedure applied to our previous example.

Example 4.4.1 We want to solve, by substitution, the integral:

$$\int e^{3x} dx$$

We set $u = 3x$. We have to be careful because there is no precise rule to choose the function $g(x)$, but we have to "try" and see if the chosen substitution allows us to carry out all the steps described above. When we have an exponential or a logarithmic function, we can try a substitution taking u as the function argument, in this case $3x$.

So, we have $du = 3dx$. If we substitute:

$$\int e^{3x} dx = (1/3) \int e^u du = (1/3)e^u + c = (1/3)e^{3x} + c$$

Now let us look at some more complicated examples.

Example 4.4.2 We want to solve the integral by substitution:

$$\int e^{2x^2} x dx$$

We set $u = 2x^2$. Hence, we have $du = 4x dx$. If we substitute:

$$\int e^{2x^2} x dx = \int e^u x du/4x = (1/4) \int e^u du = (1/4)e^{2x^2} + c$$

We now see what would have happened, if we did the same substitution for the integral:

$$\int e^{2x^2} dx = \int e^u du/4x$$

At this point, we can no longer proceed, because we have both u and x appearing in the integral. Hence, we cannot solve the integral with this substitution. Actually, we cannot solve this integral in what is called a *closed form*. That is, despite the fact that the function $f(x) = e^{2x^2}$ is continuous, we cannot find a simple function, as those studied in Chapter 1, which is the primitive of $f(x)$.

Example 4.4.3 We want to solve the following integral by substitution:

$$\int \sqrt{2x + 3} dx$$

We set $u = 2x + 3$. So, we have $du = 2dx$. If we substitute:

$$\int \sqrt{2x + 3}\, dx = (1/2) \int \sqrt{u}\, du = (1/2)\frac{u^{3/2}}{3/2} =$$

$$= (1/3)u^{3/2} + c = (1/3)(2x + 3)^{3/2} + c \tag{4.3}$$

From the above example, we see a rule for a quicker substitution:

$$\int f(x)^n f'(x)\, dx = \frac{f(x)^{n+1}}{n+1} + c \tag{4.4}$$

since we immediately see that the derivative of the function $\frac{f(x)^{n+1}}{n+1}$ is just $f(x)^n f'(x)$. Let us revisit the previous example using the formula (4.4).

Example 4.4.4 To solve the integral, using (4.4), we "reconstruct" the derivative of $f(x) = 2x + 3$, without doing the substitution explicitly:

$$\int \sqrt{2x + 3}\, dx = \frac{1}{2} \int (2x + 3)^{1/2} \cdot 2dx = \frac{1}{2}\frac{(2x + 3)^{3/2}}{3/2}$$

$$+ c = \frac{1}{3}(2x + 3)^{3/2} + c$$

We conclude this section with a result regarding the integration by substitution formula, which we have used above. The proof is immediate by looking at the chain rule for derivation.

Proposition 4.4.5 Integration by substitution formula: *Let F be a primitive of f, i.e., $F' = f$. Then:*

$$\int f(g(x))g'(x)\, dx = F(g(x)) + c$$

4.5 INTEGRATION BY PARTS

The method of integration by parts takes advantage of the Leibniz rule for the derivative of a product, studied in Chapter 2. In fact, if we have two differentiable functions f and g, we can write the derivative of their product through the Leibniz rule:

$$D[f(x)g(x)] = f'(x)g(x) + f(x)g'(x)$$

So, in terms of primitives, i.e., taking the indefinite integral for the two terms of the equality, we have:

$$f(x)g(x) = \int f'(x)g(x)dx + \int f(x)g'(x)dx$$

Hence, we obtain the *formula for integration by parts*:

$$\int [f'(x)g(x)]dx = f(x)g(x) - \int [f(x)g'(x)]dx$$

If we want to apply this formula in practice, the function we want to integrate should be the product of two functions. However, as we will see, one of the two functions could be a constant. Moreover, we should also know the integral of one of the two functions. Let us see an example.

Example 4.5.1 We want to compute the integral $\int xe^x dx$. Let us take $g(x) = x$ and $f'(x) = e^x$. Hence, we have $f(x) = e^x$. Applying the integration by parts formula:

$$\int xe^x dx = xe^x - \int e^x dx = xe^x - e^x + c$$

Let us see what happens if we choose $f'(x) = x$, $g(x) = e^x$:

$$\int xe^x dx = (x^2/2)e^x - \int e^x(x^2/2)dx$$

We obtain a more complicated integral that we are currently unable to solve. So, we see that, sometimes, we may need few attempts before choosing the correct functions to apply the formula of integration by parts.

It is also important to notice that, when applying the formula, we should get an integral which is simpler than the original one. Moreover, sometimes it is necessary to apply the formula of integration by parts a few times, obtaining simpler and simpler integrals, until we reach an integral that we can compute.

We now see an example, where it is necessary to apply the formula of integration by parts several times in order to compute the integral.

Example 4.5.2 We want to compute the integral $\int x^2 e^x dx$. Let us take, as in the previous example, $g(x) = x^2$ and $f'(x) = e^x$. Hence, we have $f(x) = e^x$. Applying the integration by parts formula:

$$\int x^2 e^x dx = x^2 e^x - \int 2x e^x dx = x^2 e^x - 2(x e^x - e^x) + c$$

where we have used the calculation seen in the previous exercise.

From this example we see that, in the formula by integration by parts, we must always choose $f'(x)$ as a function, whose primitive we know, but more importantly, as $g(x)$, a function whose derivative simplifies the calculation.

Now let us look at an example where we combine the integration by substitution technique with the formula of integration by parts.

Example 4.5.3 We want to compute the integral $\int x^2 e^x dx$. Applying the integration by parts formula:

$$\int x^2 e^{3x} dx = \frac{1}{3} e^{3x} x^2 - \int \frac{1}{3} e^{3x} 2x dx$$

Notice that the integral of e^{3x} was obtained through integration by substitution as in Example 4.4.1. Now, we apply again the integration by parts formula to the last integral:

$$\int x e^{3x} dx = \frac{1}{3} e^{3x} x - \int \frac{1}{3} e^{3x} dx = \frac{1}{3} x e^{3x} - \frac{1}{9} e^{3x}$$

Considering both of our calculations, we get:

$$\int x^2 e^{3x} dx = \frac{1}{27} e^{3x} (9x^2 - 6x + 2) + c$$

Finally, let us see an example in which we take $f(x)$ as a *constant*.

Example 4.5.4 We want to compute the integral $\int \log(x) dx$, using the formula of integration by parts. We take $f'(x) = 1$ and $g(x) = \log(x)$. Hence:

$$\int \log(x) dx = x \log(x) - \int x \cdot \frac{1}{x} dx = x \log(x) - x + c$$

We have provided a list of integrals of common functions in Appendix 4.11.

4.6 INTEGRATION OF RATIONAL FUNCTIONS

We want to compute the integral of a rational function. In other words, we want to solve an integral of the form:

$$\int \frac{p(x)}{q(x)} dx \tag{4.5}$$

where $p(x)$ and $q(x)$ are two polynomials (see Chapter 1).

If the degree of the numerator $p(x)$ is greater than or equal to the degree of the denominator $q(x)$, we can divide the numerator by the denominator obtaining a quotient $u(x)$ and a remainder $r(x)$:

$$\frac{p(x)}{q(x)} = u(x) + \frac{r(x)}{q(x)}$$

Hence, the integral (4.5) becomes:

$$\int \frac{p(x)}{q(x)} dx = \int u(x)dx + \int \frac{r(x)}{q(x)} dx$$

In the first integral, we have a polynomial, while in the second one, we have a rational function, where the numerator has a degree strictly lower than the denominator. Then, we focus on this case only. The method, that we are going to describe, always leads us to the solution of the integral and takes the name of *partial fractions method*.

Let us examine the three cases separately:

1. The denominator $q(x)$ decomposes into the product of distinct, irreducible factors of degree 1 (for example: $q(x) = x^2 - 4 = (x-2)(x+2)$).

2. The denominator $q(x)$ decomposes into the product of distinct irreducible factors of degrees 1 and 2 (for example: $q(x) = x^3 + x = x(x^2+1)$).

3. The denominator $q(x)$ decomposes into the product of factors of degrees 1 and 2 not necessarily distinct (for example: $q(x) = x^3 - x^2 = x^2(x-1)$).

Let us start with the first case, which is simpler than the others. We shall illustrate the method through various examples, leaving the general case as an exercise, since it is a straightforward application of the same procedure.

Example 4.6.1 We want to compute

$$\int \frac{1}{x^2 - 4} dx$$

We decompose the integrand function into the sum of the two functions:

$$\frac{1}{x^2 - 4} = \frac{A}{x - 2} + \frac{B}{x + 2} = \frac{Ax + 2A + Bx - 2B}{x^2 - 4}$$

We obtain the linear system:

$$\begin{cases} A + B = 0 \\ 2A - 2B = 1 \end{cases}$$

A quick calculation gives $A = 1/4$ and $B = -1/4$. Hence, we can write the integral as:

$$\int \frac{1}{x^2-4} dx = (1/4) \int \frac{1}{x-2} dx - (1/4) \int \frac{1}{x+2}$$

$$= (1/4) \log |x - 2| - (1/4) \log |x + 2| + c$$

Let us look at a more complicated case.

Example 4.6.2 We want to compute

$$\int \frac{2x - 1}{x^2 - 3x + 2} dx$$

First, we decompose the denominator:

$$x^2 - 3x + 2 = (x - 1)(x - 2)$$

Then, we decompose the integrand function into the sum:

$$\frac{2x - 1}{x^2 - 3x + 2} = \frac{A}{x - 1} + \frac{B}{x - 2} = \frac{Ax - 2A + Bx - B}{x^2 - 3x + 2}$$

We obtain the linear system:

$$\begin{cases} A + B = 2 \\ -2A - B = -1 \end{cases}$$

As solutions, we get $A = -1$ and $B = 3$. Hence, we can write the integral as:

$$\int \frac{2x-1}{x^2-3x+2} dx = - \int \frac{1}{x-1} dx + 3 \int \frac{1}{x-2} dx =$$

$$= - \log |x - 1| + 3 \log |x - 2| + c$$

Let us now look at the second case, namely, the case in which, in the decomposition of the denominator, there are irreducible factors of degree 2. For these types of integrals, we recall the derivative of the arctan function that we have computed in Chapter 2:

$$D(\arctan(x)) = \frac{1}{1 + x^2}$$

We illustrate the method of partial fractions in this second case through an example, leaving again the general case as an exercise.

Example 4.6.3 We want to compute the integral:

$$\int \frac{x + 1}{2x^3 + x} dx$$

We proceed in a similar way as we did before: we decompose the denominator into factors: $2x^3 + x = x(2x^2 + 1)$. Then, we proceed as before, but with an important difference: we take a polynomial of degree 1 as one of the numerators, rather than a constant, as we did in the previous case:

$$\frac{x + 1}{2x^3 + x} dx = \frac{A}{x} + \frac{Bx + C}{2x^2 + 1}$$

Hence, solving the linear system, we have: $A = 1$, $B = -2$, $C = 1$. We substitute in the integral:

$$\int \frac{x + 1}{2x^3 + x} dx = \int \frac{1}{x} dx + \int \frac{-2x + 1}{2x^2 + 1} dx$$

Let us focus on the second term, which we divide into two parts:

$$\int \frac{-2x + 1}{2x^2 + 1} dx = \int \frac{-2x}{2x^2 + 1} dx + \int \frac{1}{2x^2 + 1} dx$$

To solve the integral of the first part, we carry out the following substitution:

$$u = 2x^2 + 1 \implies du = 4x dx$$

from which

$$\int \frac{-2x}{2x^2 + 1} dx = -\frac{1}{2} \int \frac{1}{u} du = -\frac{1}{2} \log |2x^2 + 1| + c$$

For the second part, we use the integral of the arctan function:

$$\int \frac{1}{(x\sqrt{2})^2 + 1} dx = \frac{1}{\sqrt{2}} \arctan x\sqrt{2} + c$$

where we made the substitution: $u = x\sqrt{2}$. Hence, adding up all the integrals we have computed, we obtain:

$$\int \frac{x+1}{2x^2+1} dx = \log |x| - \frac{1}{2} \log |2x^2 + 1| + \frac{1}{\sqrt{2}} \arctan x\sqrt{2} + c$$

Finally, we come to the last case, when the denominator decomposes into irreducible factors that are not necessarily distinct. Let us see with an example how to proceed, the general case is easily obtainable and we leave it to the reader as an exercise.

Example 4.6.4 We want to compute the integral:

$$\int \frac{1}{x^2(x-2)} dx$$

We apply the method of partial fractions using, however, the following decomposition:

$$\frac{1}{x^2(x-2)} = \frac{A}{x} + \frac{B}{x^2} + \frac{C}{x-2}$$

We note that we had to include, in addition to the term $1/x^2$, also the fraction $1/x$. Solving the system, we get $A = -1/4$, $B = -1/2$, $C = 1/4$. Hence, we have:

$$\int \frac{1}{x^2(x-2)} dx = -\int \frac{1}{4x} dx - \int \frac{1}{2x^2} dx + \int \frac{1}{4(x-2)} dx =$$

$$= -\frac{1}{4} \log |x| + \frac{1}{2x} + \frac{1}{4} \log |x-2| + c$$

Since we have not treated in full generality the method of partial fractions, we invite the student to go to the section of the exercises with solutions, where we give further examples of the cases.

4.7 INTEGRATION OF TRIGONOMETRIC FUNCTIONS

In this section, we want to briefly explain how to compute integrals of the type

$$\int R(\sin x, \cos x) dx$$

where $R(\sin x, \cos x)$ is a simple function, for example, rational, in $\sin x$ and $\cos x$. To solve these integrals, we use the trigonometric formulas that express $\sin x$ and $\cos x$ as a function of $\tan(\frac{x}{2})$:

$$\sin x = \frac{2\tan(\frac{x}{2})}{1 + \tan(\frac{x}{2})^2}, \qquad \cos x = \frac{1 - \tan(\frac{x}{2})^2}{1 + \tan(\frac{x}{2})^2}, \qquad \tan x = \frac{2\tan(\frac{x}{2})}{1 - \tan(\frac{x}{2})^2}.$$

We leave these formulae as an easy check to the reader.

To solve these integrals, we need the following substitution: $t = \tan(\frac{x}{2})$. Since $x = 2\arctan t$, we immediately have that:

$$dx = \frac{2dt}{1 + t^2}$$

Once we performed this substitution, we always obtain a rational function in t. Then, we can use the method of partial fractions or another substitution. Let us see an example.

Example 4.7.1 We want to compute the integral:

$$\int \frac{1}{\sin x + \cos x + 1} dx$$

With the substitutions we have seen above, we have that:

$$= \int \frac{1}{\frac{2t}{1+t^2} + \frac{1-t^2}{1+t^2} + 1} \cdot \frac{2}{1+t^2} dt$$

$$= \int \frac{1}{1+t} dt = \log|1 + t| + c$$

With the substitution $t = \tan(\frac{x}{2})$, we get:

$$\int \frac{1}{\sin x + \cos x + 1} dx = \log|1 + \tan(\frac{x}{2})| + c$$

4.8 APPLICATIONS

In this section, we look at some examples, which show us how the concept of integral, introduced in this chapter, can be extremely useful in the applications. Let us start with an example regarding physics.

Example 4.8.1 Suppose we know the instantaneous velocity $v(t)$, expressed in meters per second, of a body in motion, as a function of time:

$$v(t) = 5t + 3$$

We want to determine the distance covered in the interval of time between 3 and 6 seconds. From Chapter 3, we know that the instantaneous velocity, being the rate of change of the distance traveled in time, is equal to the derivative of the distance covered $s(t)$, with respect to the time. Hence, we have:

$$s'(t) = v(t)$$

To find the distance traveled in the given interval of time, we need to compute the following definite integral:

$$\int_3^6 (5t + 3)dt = \left[5\frac{t^2}{2} + 3t\right]_3^6 = \frac{153}{2}$$

Thus, between 3 and 6 seconds, the body will travel 76.5 m.

Now we see an example from biology.

Example 4.8.2 Suppose, in a certain environment, a population of individuals grows with a growth rate expressed by the function

$$C(t) = \frac{4000}{\sqrt{t}}$$

where t represents the time measured in days.

This is not a Malthusian law; in fact, such law is exponential $N(t) = N_0 e^{\lambda t}$, and the growth rate is proportional to size of the population, as we will study in later chapters. This could happen, for example, if the population reproduces initially according to Malthus law, but with limited resources available. Then, the expression of the population size as the function of time would no longer be an exponential. We will see more concrete examples later, such as Bertalanffy's law regulating tumor growth, that we will be studied with more sophisticated tools.

Suppose, for $t = 1$, that is, on the first day, we count $25,000$ individuals in the population. We want to compute the population size $P(t)$ as time changes.

The population growth rate represents the change of $P(t)$ over time, that is, its derivative. Hence, if we want to determine $P(t)$, we need to take the integral of the growth rate:

$$\int C(t)dt = \int \frac{4000}{\sqrt{t}}dt = \int 4000t^{-1/2}dt = 4000(2t^{1/2})+c = 8000\sqrt{t}+c.$$

To find the value of the constant of integration c, we use the information we have at time $t = 1$: $P(1) = 25,000$. Hence:

$$25,000 = 8000 + c$$

obtaining $c = 17,000$. So, the variation of the population over time is given by the law:

$$P(t) = 8000\sqrt{t} + 17,000$$

Now we want to establish how many days are necessary to have the population reach 97,000 individuals. We proceed by setting $P(t) = 97,000$ and then solve to obtain the value of t:

$$8000\sqrt{t} + 17,000 = 97,000$$

We have $\sqrt{t} = 10$ and therefore $t = 100$. So, the population will reach $97,000$ individuals after 100 days.

We will see in the following chapters more general and effective methods to determine the solution of similar problems through the theory of differential equations.

4.9   EXERCISES WITH SOLUTIONS

4.9.1 Compute the area between the graph of function $f(x) = \log(x - 1)$ and the x-axis of x between $3/2$ and 3.

Solution. We observe that the function changes sign for $x = 2$: $f(x) \leq 0$ for $3/2 \leq x \leq 2$ and $f(x) \geq 0$ for $2 \leq x \leq 3$. Hence, the area that we have to determine is given by:

$$\text{Area} = -\int_{3/2}^{2} \log(x - 1)dx + \int_{2}^{3} \log(x - 1)dx$$

We then proceed with the calculation of the indefinite integral by substitution, setting $u = x - 1$:

$$\int \log(x-1)dx = (x-1)\log|x-1| - (x-1) + c$$

Hence, the requested area is given by:

$$-[(x-1)\log|x-1| - (x-1)]_{3/2}^2 + [(x-1)\log|x-1| - (x-1)]_2^3 =$$

$$= 0.153 + 0.386 = 0.539$$

4.9.2 Compute the area of the flat region bounded by the graphs of functions $f(x) = x^2$ and $g(x) = \sqrt{x}$, for x in the interval $[0, 1]$.

Solution. First of all, we observe that we are dealing with positive functions in the given interval. Furthermore, we see graphically that $x^2 \le \sqrt{x}$ (see Chapter 1). Hence, the required area is the difference between the area between the curve $y = \sqrt{x}$ and the x-axis and the area between the curve $y = x^2$ and the x-axis:

$$A = \int_0^1 (\sqrt{x} - x^2)dx = \int_0^1 \sqrt{x}dx - \int_0^1 x^2 dx =$$

$$= \left[\frac{2}{3}\sqrt{x^3}\right]_0^1 - \left[\frac{x^3}{3}\right]_0^1 = 2/3 - 1/3 = 1/3.$$

4.9.3 Compute the integral $\int \frac{x^2 + 4x - 1}{\sqrt{x}} dx$.

Solution. First, we divide each term of the numerator by the denominator:

$$\int \frac{x^2 + 4x - 1}{\sqrt{x}} dx = \int \left(x^{3/2} + 4x^{1/2} - x^{-1/2}\right) dx$$

Applying the formula for integrating the powers of x, we get immediately:

$$\int \frac{x^2 + 4x - 1}{\sqrt{x}} dx = \frac{2\sqrt{x^5}}{5} + \frac{8\sqrt{x^3}}{3} - 2\sqrt{x} + c$$

4.9.4 Compute the integral $\int \sin(2x + 3)\, dx$.

Solution. We set $u = 2x + 3$, hence $du = 2dx$. If we substitute:

$$\int \sin(2x+3)\, dx = (1/2)\int \sin u\, du = -(1/2)\cos(2x+3) + c$$

4.9.5 Compute the integral $\int \frac{\log(x)+\log^2(x)}{x}\,dx$.

Solution. We set $\log x = t$, from which $x = e^t$ and substitute :

$$\int \frac{\log(x)+\log^2(x)}{x}dx = \int \frac{t+t^2}{e^t}e^t dt = \int (t+t^2)dt = \frac{t^2}{2}+\frac{t^3}{3}+c$$

Substituting again we obtain:

$$\int \frac{\log x + \log^2(x)}{x}dx = \frac{\log^2(x)}{2}+\frac{\log^3(x)}{3}+c$$

4.9.6 Compute the integral:

$$-\int \frac{2}{2+e^x}\,dx$$

Solution. We proceed by substitution by setting $u = e^x$, hence $du = e^x dx$. Hence:

$$-\int \frac{2dx}{2+e^x} = -\int \frac{2du}{u(2+u)}$$

We use the method of partial fractions

$$-\frac{2}{u(u+2)} = \frac{A}{u}+\frac{B}{u+2}$$

obtaining $A = -1$, $B = 1$.

$$-\int \frac{2du}{u(2+u)} = -\int \frac{1}{u}du + \int \frac{1}{u+2}du = -\log|u| + \log|u+2| + c$$

Hence, substituting $u = e^x$:

$$-\int \frac{2}{2+e^x}\,dx = \log \frac{e^x+2}{e^x}+c$$

4.9.7 Compute the integral $\int \frac{4x+5}{2x^2-x-1}dx$

Solution. By the method of partial fractions, we can immediately write:

$$\frac{4x+5}{2x^2-x-1} = \frac{4x+5}{(2x+1)(x-1)} = \frac{A}{2x+1}+\frac{B}{x-1}$$

So, we have $4x+5 = A(x-1)+B(2x+1)$, hence $C = -2$ and $D = 3$.
Substituting in the integral:

$$\int \frac{4x+5}{2x^2-x-1}dx = \int \frac{-2}{2x+1}dx + \int \frac{3}{x-1}dx = -\log|2x+1|+3\log|x-1|+c$$

4.9.8 Compute the integral $\int \frac{x+6}{x^2-4x+8} dx$.

Solution. We note that the denominator has no real roots. We then write the function as sum of two terms, where in the numerator of the first term we reconstruct the derivative of the denominator, so that we can proceed with a substitution.

$$\frac{x+6}{x^2-4x+8} = \frac{1}{2}\frac{2x+12}{x^2-4x+8} = \frac{1}{2}\frac{2x-4+4+12}{x^2-4x+8} =$$

$$= \frac{1}{2}\frac{2x-4}{x^2-4x+8} + \frac{1}{2}\frac{16}{x^2-4x+8}$$

The given integral becomes the sum of two integrals:

$$\int \frac{x+6}{x^2-4x+8} dx = \frac{1}{2}\int \frac{2x-4}{x^2-4x+8} dx + \int \frac{8}{x^2-4x+8} dx$$

Now we do a simple substitution $u = x^2 - 4x + 8$ to solve the first integral:

$$\frac{1}{2}\int \frac{2x-4}{x^2-4x+8} dx = \frac{1}{2}\log(x^2-4x+8) + c$$

For the second integral, we want to reconstruct the integral of the arctan. We observe that in order to have:

$$x^2 - 4x + 8 = (x-2)^2 + k$$

we need $k = 4$. We obtain:

$$\int \frac{8}{x^2-4x+8} dx = \int \frac{8}{(x-2)^2+4} dx$$

We substitute $x - 2 = t$ and $dx = dt$:

$$\int \frac{8}{t^2+4} dt = 8\int \frac{1}{t^2+4} dt = 8\frac{1}{\sqrt{4}}\arctan(\sqrt{\frac{1}{4}}t) + c = 4\arctan(\frac{t}{2}) + c$$

Now substituting t for $x - 2$, we obtain:

$$\int \frac{x+6}{x^2-4x+8} dx = \frac{1}{2}\log(x^2-4x+8) + 4\arctan(\frac{x-2}{2}) + c$$

4.9.9 Compute the integral $\int \frac{x+1}{9x^2-6x+1}dx$.

Solution. We observe that:

$$\frac{x+1}{9x^2-6x+1} = \frac{x+1}{(3x-1)^2}$$

We need to find A and B such that :

$$\frac{x+1}{(3x-1)^2} = \frac{A}{(3x-1)} + \frac{B}{(3x-1)^2}$$

A quick calculation, as explained above, allows us to find: $A = \frac{1}{3}$ and $B = \frac{4}{3}$. Then, the given integral decomposes into the sum of the following two integrals:

$$\int \frac{x+1}{9x^2-6x+1}dx = \int \frac{1/3}{(3x-1)}dx + \int \frac{4/3}{(3x-1)^2}dx$$

which we can easily solve by substitution (put $u = 3x - 1$) finding as final result:

$$\int \frac{x+1}{9x^2-6x+1}dx = \frac{1}{9}\log|3x-1| - \frac{4}{27x-9} + c$$

4.10 SUGGESTED EXERCISES

4.10.1 Compute the following indefinite integrals by substitution:

1. $\int x\sqrt{2x^2+1}dx$

2. $\int \frac{x}{\sqrt{1-3x^2}}dx$

3. $\int x\cos(x^2+1)dx$

4. $\int x^2 e^{1-x^3}dx$

5. $\int \frac{3x^2-2}{4-2x+x^3}dx$

6. $\int x\sin(2+x^2)\,dx$

4.10.2 Compute the following definite integrals by substitution:

1. $\int_0^1 \sqrt{2x+1}\,dx$

2. $\int_0^{\pi/2} \cos(x)\sin^2(x)\,dx$

3. $\int_0^{\pi/2} \frac{\cos(x)}{1+\sin(x)}\,dx$

4. $\int_1^2 \frac{\log(2x)}{x}\,dx$

5. $\int_0^1 x^2 e^{1-x^3}\,dx$

6. $\int_0^2 x(x+1)^4\,dx$

4.10.3 Compute the following integrals by parts, possibly using also the method for substitution, if necessary:

1. $\int xe^{4x+1}\,dx$

2. $\int x\cos(2x-1)\,dx$

3. $\int x^2 \sin(2x)\,dx$

4. $\int x\log(x)\,dx$

5. $\int \sqrt{x}\log(x)\,dx$

4.10.4 Compute the following integrals:

1. $\int \frac{2x}{\sqrt{2x+1}}\,dx$

2. $\int \cos\sqrt{x}\,dx$ [Hint: substitute $u = \sqrt{x}$]

3. $\int \tan(x)\,dx$ [Hint: by substitution]

4. $\int \sqrt{x}e^{\sqrt{x}}\,dx$ [Hint: by substitution]

5. $\int \cos\log(x)\,dx$

6. $\int x^2 \log(2x)\,dx$

7. $\int x(\log(2x))^2\,dx$

4.10.5 Compute the following integrals:

1. $\int \frac{x}{x^2-9} dx$

2. $\int \frac{x-1}{x^2-x} dx$

3. $\int \frac{2x-3}{x^2-5x+6} dx$

4. $\int \frac{x^2}{x^2+1} dx$

5. $\int \frac{1}{(x-2)(x-4)} dx$

6. $\int \frac{1}{(x^2-x-2)} dx$

7. $\int \frac{1}{x^2+9} dx$

8. $\int \frac{x}{x^2(x-3)} dx$

4.10.6 For each of the following functions, compute the area between the graph and the x-axis in the given interval.

1.
$$f(x) = \frac{x-1}{x-6}, \quad [0,1]$$

2.
$$f(x) = \frac{x-2}{x+5}, \quad [1,5]$$

3.
$$f(x) = \frac{x-1}{x^2-x-6}, \quad [-1,0]$$

4.
$$f(x) = e^{-x} - e^{-3x}, \quad [0,1]$$

5.
$$f(x) = \frac{1}{x^2+x-2}, \quad [-1/2, 1/2]$$

4.10.7 Determine the area between $y = f(x)$ and the x-axis of x for the following functions in the given interval:

1. $f(x) = xe^{2x^2-1}$ in the interval $[-1,1]$.

2. $f(x) = x\cos(x)$ in the interval $[-\pi, \pi]$.

3. $f(x) = (12x - 6)/(x^2 - 3)$ in the interval $[-1, 0]$.

4. $f(x) = \sin\sqrt{x}$ in the interval $[0, \pi^2]$.

5. $f(x) = \log(x)$ in the interval $[1/2, 3/2]$.

4.10.8 In the first weeks of sprouting, a bean sprout has a growth rate that follows the law:
$$C(t) = 3t^2 + t$$
where t is expressed in weeks and the growth is expressed in mm/week.

Determine after how many days the sprout has reached the height of 10 mm, knowing that at time $t = 0$ germination begins.

4.10.9 Suppose the function $f(t) = e^t + 2$ describes the variation of a certain population in time measured in hours and that, at time $t = 0$, the population counts 150 individuals. Determine how many individuals the population counts after 4 hours.

4.11 APPENDIX: INDEFINITE INTEGRALS

1. $\int k dx = kx + c$

2. $\int x^n dx = \frac{x^{n+1}}{n+1} + c$ with $n \neq -1$

3. $\int \frac{1}{x} dx = \log|x| + c$

4. $\int e^x dx = e^x + c$

5. $\int \sin x dx = -\cos x + c$

6. $\int \cos x dx = \sin x + c$

7. $\int \frac{1}{1+x^2} dx = \arctan x + c$

8. $\int \frac{1}{\sqrt{1-x^2}} dx = \arcsin x + c$

9. $\int -\frac{1}{\sqrt{1-x^2}} dx = \arccos x + c$

4.12 APPENDIX: THEOREMS ON INTEGRAL CALCULUS

In this section, we give the proofs of some fundamental results of the theory of integration of this chapter.

We start with the theorem establishing the integrability of continuous functions on a closed interval: this theorem is of extreme importance since, as we have seen, all functions of interest in applications are continuous.

Theorem 4.12.1 *A function f continuous in $I = [a, b]$ is integrable.*

Proof. Since f is continuous, we have that f is bounded on $[a, b]$ by the Weierstrass Theorem. We also have that f is uniformly bounded on $[a, b]$.

In the case of f derivable, as for all functions of interest to applications, this is true. This means that:
$\forall \epsilon > 0$ fixed, $\exists \delta > 0$ such that if $y_1, y_2 \in [a, b]$ and $|y_1 - y_2| < \delta$ then $|f(y_1) - f(y_2)| < \epsilon$.
Consider a partition $P = \{x_0, x_1, \ldots, x_n\}$ of $[a, b]$ such that each interval $[x_{k-1}, x_k]$ has a length smaller than δ for $k = 1, \ldots, n$.
We apply Weierstrass' Theorem on each of these subintervals: for each $k = 1, \ldots, n$ there are points $\zeta_k, \eta_k \in [x_{k-1}, x_k]$ such that:

$$f(\zeta_k) = m_k = \min_{x \in [x_{k-1}, x_k]} f(x)$$

and

$$f(\eta_k) = M_k = \max_{x \in [x_{k-1}, x_k]} f(x)$$

Since $|\eta_k - \zeta_k| < \delta$, we have: $M_k - m_k = f(\eta_k) - f(\zeta_k) < \epsilon$.
Let now $h_\epsilon \geq f$ and $g_\epsilon \leq f$, piecewise constant functions, defined as follows:

$$h_\epsilon(x) = \begin{cases} M_k & x \in (x_{k-1}, x_k] \\ f(a) & x = a \end{cases}$$

and

$$g_\epsilon(x) = \left\{ \begin{array}{ll} m_k & x \in (x_{k-1}, x_k] \\ f(a) & x = a \end{array} \right.$$

For each $x \in [a, b]$ the result is $h_\epsilon(x) - g_\epsilon(x) < \epsilon$ and then:

$$\int_I h_\epsilon(x)dx - \int_I g_\epsilon(x)dx = \int_I (h_\epsilon(x) - g_\epsilon(x))dx < \int_I \epsilon dx = \epsilon(a - b).$$

Since ϵ is arbitrary, if $\epsilon \to 0$ the desired result is obtained, since:

$$\int_I g_\epsilon(x)dx \leq \int_I f(x)dx \leq \int_I h_\epsilon(x)dx$$

□

Let us introduce the Fundamental Theorem of Calculus which establishes a link between two important operations: derivation and integration.

Theorem 4.12.2 Fundamental Theorem of Calculus, Part I: *Let f be a continuous function on $[a, b]$ and let F be a primitive of f, i.e., a function such that $F'(x) = f(x)$ on $[a, b]$. Then,*

$$\int_a^b f(x)dx = F(b) - F(a)$$

This result allows us to compute the definite integral of f without taking limits, when we have a primitive F of f.

Proof. Consider an arbitrary partition of $[a, b]$:

$$a = x_0 < x_1 < x_2 < \cdots < x_N = b$$

We write the difference $F(b) - F(a)$ as the sum of the differences of $F(x)$ in the subintervals $[x_{i-1}, x_i]$:

$$F(b) - F(a) = F(x_N) - F(x_0) =$$

$$= (F(x_1) - F(x_0)) + (F(x_2) - F(x_1)) + \cdots + (F(x_N) - F(x_{N-1})) =$$

$$= \sum_{i=1}^N (F(x_i) - F(x_{i-1})) \tag{4.6}$$

Now, let us use the Mean Value Theorem or Lagrange Theorem: in each subinterval $[x_{i-1}, x_i]$, there exists a point c_i^*, such that:

$$\frac{F(x_i) - F(x_{i-1})}{x_i - x_{i-1}} = F'(c_i^*) = f(c_i^*)$$

So, $F(x_i) - F(x_{i-1}) = f(c_i^*)(x_i - x_{i-1})$ and substituting in the equation (4.6):

$$F(b) - F(a) = \sum_{i=1}^{N} f(c_i^*)(x_i - x_{i-1}) = \sum_{i=1}^{N} f(c_i^*)\Delta_i \qquad (4.7)$$

The latter is the approximation in Riemann sum of the integral $\int_a^b f(x)dx$. Since f is an integrable function, the sums approximate the integral $\int_a^b f(x)dx$, when $\Delta = \max_i \Delta_i$ tends to zero. On the contrary, the Riemann sums, for (4.7), are equal to $F(b) - F(a)$, due to the choice of c_i^*'s. Then, we have our result:

$$F(b) - F(a) = \int_a^b f(x)dx$$

$\square$

The second part of the Fundamental Theorem of Calculus tells us that we can use the definite integral of f to compute a primitive of f.
We introduce the *signed* area function of the graph of f in the interval $[a, x]$:

$$A(x) = \int_a^x f(t)dt$$

Theorem 4.12.3 Fundamental Theorem of Calculus, Part II:
Let f be a continuous function in $[a, b]$ and let $A(x) = \int_a^x f(t)dt$ be a primitive of f, equivalently $A'(x) = f(x)$. Then:

$$\frac{d}{dx}\int_a^x f(t)dt = f(x)$$

Also, $A(x)$ satisfies the initial condition $A(a) = 0$

Proof. To simplify the proof, suppose f is monotone increasing in $[a, b]$, i.e., $f(x_1) \leq f(x_2)$ for $x_1 < x_2$. We compute $A'(x)$ as the limit of the difference quotient, and we begin to study the numerator of the difference quotient. By the addition property on the domain of the definite integral, we have:

$$A(x+h) - A(x) = \int_a^{x+h} f(t)dt - \int_a^{x} f(t)dt = \int_x^{x+h} f(t)dt$$

Suppose $h > 0$, then the numerator of the difference quotient is equal to the area between the graph of f and the x-axis between x to $x+h$.

This area contains and is contained in two rectangles of height, respectively, $f(x)$ and $f(x+h)$, both with base length h. Hence:

$$h \cdot f(x) \leq A(x+h) - A(x) \leq h \cdot f(x+h)$$

Dividing by h we get:

$$f(x) \leq \frac{A(x+h) - A(x)}{h} \leq f(x+h)$$

Now we apply the Comparison Theorem and the continuity of f, which implies that $\lim_{h\to 0} f(x+h) = f(x)$ and that $\lim_{h\to 0} f(x) = f(x)$. We conclude that:

$$\lim_{h\to 0^+} \frac{A(x+h) - A(x)}{h} = f(x)$$

If $h < 0$, just consider $A(x) - A(x+h)$ as numerator of the difference quotient and $(-h) > 0$ in place of h, if we want the same interpretation in terms of areas. At this point, the proof follows in a similar way to the case $h > 0$. Hence, $A'(x) = f(x)$. □

CHAPTER 5

First-Order Differential Equations

5.1 FIRST-ORDER EQUATIONS

A differential equation is an equation in which we have, as unknown, a function together with its derivatives. For example, consider the differential equation:

$$y'(x) = x + 1$$

The solution is given by the function $y(x) = x^2/2 + x + C$, where C is an arbitrary constant, that is, $y(x)$ is a solution for every real value of C. To get the solution, we took the integral of the function $x + 1$, that is, we found a function $y = y(x)$, whose derivative is $x + 1$. We call *general solution* of a differential equation, the most general form that a solution can take, while we call *particular solution* just one of its solutions. For example, for the equation $y' = x + 1$, we have general solution $y(x) = x^2/2 + x + C$, while a particular solution is given, for example, by setting $C = 2$: $y(x) = x^2/2 + x + 2$. It is clear that, determining the general solution, i.e., an expression that gives all the solutions of a given differential equation at once, is more difficult than determining a particular solution and it is not always possible. However, we will see that, in some cases of particular interest, we can always find the general solution.

We say that a differential equation is of *first order*, if only the first derivative of the function, that we want to determine, appears in the equation. For example, the equation $y' = x + 1$ is of first order,

DOI: 10.1201/9781003343288-5

while the equation $y'' = y$ is of the second order. We also speak of *ordinary* differential equations, where the word "ordinary" refers to the fact that the unknown function depends on one variable only. If an equation contains the derivatives with respect to more than just one variable, we call it a *partial differential equation*, but we shall not deal with such a case.

A differential equation of first order is said to be *in normal form*, if:

$$y'(x) = f(x, y(x))$$

These differential equations are of particular importance, as we will see in the next section.

5.2 THE CAUCHY PROBLEM

Modeling a biological or, more generally, a dynamical system often requires to solve an initial value problem; in other words, we want to find the time evolution of a certain function knowing an initial condition and an equation for the function and its derivative. Let us see how to express mathematically this question.

Definition 5.2.1 A *initial value problem* or a *Cauchy problem* is a first-order ordinary differential equation in normal form, $y(t)' = f(t, y(t))$, together with an *initial condition* $y(t_0) = y_0$:

$$\begin{cases} y'(t) = f(t, y(t)) \\ y(t_0) = y_0 \end{cases}$$

where t_0 is in the domain of the solution $y(t)$.

We now state, without proof, the existence and uniqueness theorem for the solution of an initial value problem, provided that some regularity conditions are satisfied, which in fact we practically always have in the applied sciences. This theorem states that, under suitable hypotheses, the Cauchy problem always admits solution and that such solution is unique. For a proof of this important result, we refer the reader to [1] and [3].

Theorem 5.2.2 Cauchy's theorem: *Let $f(t, y)$ be a function whose derivatives with respect to t and y (i.e. assuming the other variable is* *a constant) are continuous in a rectangle R in the (t, y) plane. Then, the Cauchy problem:*

$$\begin{cases} y'(t) = f(t, y(t)) \\ y(t_0) = y_0 \end{cases}$$

with $(t_0, y_0) \in R$, admits a solution and the solution $y = y(t)$ is unique.

Let us see an example.

Example 5.2.3 We want to solve the Cauchy problem:

$$\begin{cases} y'(t) = t + 1 \\ y(0) = 2 \end{cases}$$

The general solution of $y'(t) = t + 1$ is given by $y(t) = t^2/2 + t + C$. To determine the constant C, we impose the initial condition $y(0) = 2$, obtaining the particular solution $y(t) = t^2/2 + t + 2$, which is the (unique) solution of the given Cauchy problem.

Given a Cauchy problem, we define the *interval of existence*, as the largest interval, where we can define the solution. For example, the solution of the previous example exists for all real numbers, i.e., for $t \in (-\infty, \infty)$. Hence, $(-\infty, \infty)$ is its interval of existence.

Let us see another example.

Example 5.2.4 We want to solve the Cauchy problem:

$$\begin{cases} y'(t) = 1/t \\ y(-1) = 3 \end{cases}$$

Clearly, we need to impose $t \neq 0$. The general solution of $y'(t) = 1/t$ is $y(t) = \log|t| + C$. If we impose the initial condition, we get $y(-1) = \log|-1| + C = 3$. Hence, the solution of the Cauchy problem is given by: $y(t) = \log|t| + 3$. We observe that the interval of existence of the solution is $(-\infty, 0)$. Hence, we can omit the absolute value and write the solution as:

$$y(t) = \log(-t) + 3, \qquad t \in (-\infty, 0).$$

5.3 DIRECTION FIELD

The field of directions of a first-order differential equation in normal form allows us to have a qualitative understanding of the graphs of the solutions, without actually solving the equation. Hence, it is a very valuable tool in applied sciences.

Suppose we have a differential equation:

$$y' = f(x, y)$$

We can interpret this equation as the expression of the slope of the tangent line to the graph of the solution $y = y(x)$, at a generic point (x, y), in terms of the function $f(x, y)$.

For example, the equation:

$$y'(x) = 2x$$

tells us that the slope of the tangent line to $y = x^2 + C$, the solution of the given differential equation, is $2x$ at the point (x, y).

An easy way to visualize such important geometric interpretation is to draw, for a certain number of equally spaced points, a line segment corresponding to a portion of the tangent line at that point to the curve $y = y(x)$, the solution of the given differential equation. The set of such line segments is called *direction field* or *slope field*.

Let us see the field of the directions of $y'(x) = 2x$:

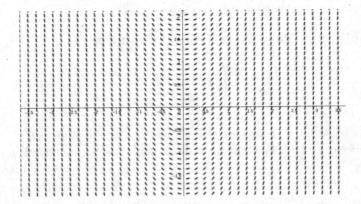

We can see that the solutions are parabolas, without actually solving the differential equation.

Let us look at another instructive example.

Example 5.3.1 Consider the differential equation $dy/dx = x^2 - x - 2$ and draw its direction field:

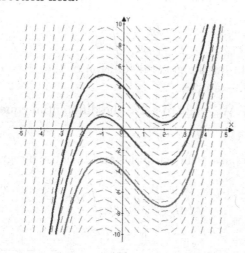

The curves represent the solutions $(x^3/3) - (x^2/2) - 2x + 4$, $(x^3/3) - (x^2/2) - 2x$, $(x^3/3) - (x^2/2) - 2x - 4$.

We see that the field of directions allows us to view an approximation of the graphs of the solutions of a first-order differential equation without solving it. We start from a point, and "following" the field of directions, we can graphically construct a curve tangent to the line segments at each point. This curve will approximate the solution of the initial value problem corresponding to the given differential equation.

5.4 SEPARABLE EQUATIONS

The method of separation of variables allows us to solve quickly the separable equations, as well as the Cauchy problems involving them.

Suppose we can write a differential equation in the form:

$$\frac{dy}{dx} = g(x)f(y) \tag{5.1}$$

Then, we say that the equation is *separable*.

We now describe the method of solution in practice together with some examples. We will discuss this procedure, in a mathematically rigorous way, at the end of this section.

We *separate the variables* in (5.1), that is, we divide by $f(y)$, so that each variable appears only on one side of the equation. Then, we take the integral:

$$\int \frac{dy}{f(y)} = \int g(x)dx \qquad (5.2)$$

We can, then, obtain the general solution of the given equation.

Let us see an example.

Example 5.4.1 We want to solve the differential equation:

$$y'(t) = -y^2 t$$

We proceed to separate the variables and take the integral, as described above:

$$\int \frac{1}{y^2} dy = - \int t \, dt$$

We get:

$$-\frac{1}{y} = -t^2/2 + c \quad \Longrightarrow \quad y = -\frac{1}{c - t^2/2}$$

Naturally, it is necessary to set $y \neq 0$ and discuss the domain for the solution.

Now we look at an example of an initial value problem.

Example 5.4.2 Let us consider the Cauchy problem:

$$\begin{cases} y'(t) = -y^2 t \\ y(0) = 4 \end{cases}$$

We first proceed to solve the differential equation, finding the general solution, that depends on a parameter c:

$$y = -\frac{1}{c - t^2/2}$$

Then, we impose the initial condition $y(0) = 4$; this allows us to determine a value for c:

$$y(0) = -\frac{1}{c} = 4 \quad \Longrightarrow \quad c = -1/4$$

Hence, the solution of the Cauchy problem is:

$$y(t) = \frac{4}{1 + 2t^2}$$

The interval of existence of the solution is given by the real numbers.

We conclude with a remark on the method of separation of variables.

Observation 5.4.3 We want to give a more formal explanation of the method, which we have applied in the previous examples, to solve separable differential equations.

Suppose we have a separable differential equation:

$$\frac{dy}{dx} = g(x)f(y) \tag{5.3}$$

We can write:

$$\frac{1}{f(y(x))}y'(x) = g(x), \qquad f(y(x)) \neq 0$$

Now, let us take the integral of both sides of the equation:

$$\int \frac{1}{f(y(x))}y'(x)dx = \int g(x)dx$$

and then we substitute: $u = y(x)$, $du = y'(x)dx$:

$$\int \frac{1}{f(u)}du = \int g(x)dx$$

We obtain the formula (5.2), where $u = y$.

5.5 NEWTON'S LAW OF COOLING

Newton's law of cooling states that the rate of change of the temperature of a body is directly proportional to the difference of temperature between the body and the ambient temperature. We are assuming that there are no sudden variations of temperature and that the nature of the heat transfer mechanism remains the same during the whole heat transfer process. Furthermore, we also assume that the ambient temperature is not affected by the body cooling down. Such assumptions are justified in the most common applied problems. Let us see how to set up the differential equation modeling this system. We recall that the rate of change of a function is given by its derivative. We can then write, most immediately, the initial value problem:

$$\begin{cases} \frac{dT(t)}{dt} = -k\left(T_A - T(t)\right), & k > 0 \\ T(0) = T_0 \end{cases}$$

where T_A is the ambient temperature, T_0 is the initial temperature and k is a constant that depends on the medium in which the transfer of heat takes place. We will see how it is possible, in some cases, to determine k.

Let us solve this equation, using the method of separation of variables:

$$\int \frac{dT}{T_A - T} = -\int k\,dt$$

from which:

$$\log(T_A - T) = -kt + c \quad \Longrightarrow \quad T - T_A = e^{-kt+c} = Ce^{-kt}, \qquad C = e^c$$

Notice that we require no conditions on the argument of the logarithm, since $T(t) > T_A$ by hypothesis, because the body is cooling. We now impose the initial condition $T(0) = T_0$, obtaining:

$$T_0 - T_A = C$$

Hence:

$$T(t) = T_A + (T_0 - T_A)e^{-kt}.$$

We observe that the interval of existence of the solution is given by all real values of t, and hence, we can derive conclusions on the body temperature even for negative times, i.e., before the initial time $t = 0$.

Now let us see an example of application of Newton's cooling law.

Example 5.5.1 A murder victim is found at midnight with a body temperature of 31°C (degree Celsius). After an hour, the coroner measures again the temperature, which has dropped to 29°C. If the ambient temperature is 21°C, we want to establish the time of death. We set up the differential equation with the available data:

$$T(t) = 21 + (31 - 21)e^{-kt}$$

We used, as the initial temperature, the one measured at the discovery of the body. To determine k, we use the fact that, after one hour, we have a temperature of 29°C:

$$T(1) = 21 + 10e^{-k} = 29$$

obtaining $k = 0.22$. So, we have:

$$T(t) = 21 + 10e^{-0.22\,t}$$

We now proceed to compute the time since the murder, assuming that, initially, the temperature of the body is 37°C:

$$T(t_d) = 21 + 10e^{-0.22\,t_d} = 37$$

Hence, we have $t_d = -2.14$. So, the murder was committed about 2 hours earlier than the discovery of the body, that is, around 10 pm.

5.6 LINEAR EQUATIONS

First-order linear differential equations are very important in modeling problems for applied sciences, as we shall see in later sections, when we will study the mixing problems and Malthusian laws. We can solve such equations through the use of an *integrating factor*, also called a *factor of integration*. This is a technique that is also used to solve other types of differential equations, which we do not treat here.

Definition 5.6.1 We define *linear equation* an equation of the type:

$$y'(x) = a(x)y(x) + f(x)$$

It may come associated with a Cauchy problem:

$$y'(x) = a(x)y(x) + f(x), \qquad y(x_0) = y_0$$

which, under suitable regularity conditions (see previous sections), has a unique solution.

To find the general solution, we make use of an *integrating factor*.

Let $u = u(x)$ be the integrating factor and multiply the solution by $u(x)$, which at the moment is unknown. If we compute the derivative, we get:

$$(u(x)y(x))' = u'(x)y(x) + u(x)y'(x) = u'(x)y(x) + u(x)[a(x)y(x) + f(x)]$$
$$(5.4)$$

We assume: $u(x)' = -a(x)u(x)$.

This assumption simplifies the problem and leads us to the solution. We can substitute in (5.4) the expression of the integrating factor:

$$(u(x)y(x))' = u'(x)y(x) + u(x)[a(x)y(x) + f(x)] =$$

$$-a(x)u(x)y(x) + u(x)a(x)y(x) + u(x)f(x) = u(x)f(x),$$

So $(uy)' = uf$, hence:

$$y(x) = \frac{1}{u(x)} \left[\int u(x) f(x) dx + C \right]$$

It is easy to check that $u'(x) = -a(x)u(x)$ has a solution and it is given by $u(x) = e^{-\int a(x) dx}$, as it is a separable differential equation. Hence, the solution $y = y(x)$ is given by the formula:

$$y(x) = e^{\int a(x) dx} \left[\int \left(e^{-\int a(x) dx} \right) f(x) dx + C \right] \tag{5.5}$$

We advice the student not to memorize the formula, but to understand how the integrating factor is used to get the solution, so that the student can write it through reasoning.

Let us see an example.

Example 5.6.2 Consider the following differential equation:

$$\begin{cases} y' = x - y \\ y(2) = 5 \end{cases} \tag{5.6}$$

We see that it is a linear equation, where $a(x) = -1$ and $f(x) = x$. Hence, we can immediately give the solution via the formula (5.5). Since $u(x) = e^x$:

$$y(x) = e^{-x} \left[\int e^x x dx + C \right]$$

Integrating by parts, we get:

$$y(x) = e^{-x} \left[e^x x - e^x + C \right] = x - 1 + Ce^{-x}$$

We now impose the initial condition to find C:

$$y(2) = 2 - 1 + Ce^{-2} = 5, \qquad \Longrightarrow C = 4e^2$$

We obtain

$$y = 4 \cdot e^{(2-x)} + x - 1$$

5.7 MIXING PROBLEMS

We want to show how the modeling, we obtain via differential linear equations, allow us to solve simple mixing problems. Let us see some of examples.

Example 5.7.1 A tank holds initially 100 L of fresh water. We pour into the tank, at a rate of 4 L/min, a mixture containing 2 g of salt per liter and the mixture leaves the tank at a rate of 4 L/min. We want to determine the amount of salt in the tank, as time changes.

We denote with $x(t)$ the amount of salt in the tank at time t. Clearly, $x(0) = 0$, because initially there is no salt inside the tank. The rate of change of the amount of salt depends on how much salt we put in the tank and how much salt we take out from the tank at every minute. Hence:

$$\frac{dx}{dt} = 4\text{L/min} \cdot 2\text{g/L} - 4\text{L/min} \cdot x(t)/100\text{L}$$

We set up the Cauchy problem:

$$\begin{cases} x'(t) = 8 - \frac{4x(t)}{100} \\ x(0) = 0 \end{cases} \tag{5.7}$$

We have a linear differential equation with $a(t) = -4/100$ and $f(t) = 8$. We proceed with the formula for the solution (5.5):

$$x(t) = e^{-4t/100}\left[\int 8e^{4t/100}dt + C\right] = 200 + Ce^{-t/25}$$

If we impose the initial condition $x(0) = 0$, we can determine the constant C:

$$200 - Ce^0 = 0, \qquad C = 200$$

So:

$$x(t) = 200 - 200e^{-t/25}$$

Now suppose that we want to compute the salt in the tank after an hour and a half:

$$x(90) = 200 - 200e^{-90/25} \cong 194.54$$

After an hour and a half, we have 194.54 g of salt in the tank.

Let us look at another more complicated example.

Example 5.7.2 A tank is filled with 200 L of fresh water. We pour into the tank, at a rate of 4 L/min, a mixture containing 1 g of salt per liter and the mixture leaves the tank at a rate of 2 L/min. We want to determine the amount of salt in the tank, as time varies, and how much salt we have in the tank after two and a half hours.

We proceed as in the previous example, denoting with $x(t)$ the amount of salt in the tank at time t, hence $x(0) = 0$. As before, the rate of change depends on how much salt we pour in the tank and how much salt we take out every minute.

$$\frac{dx}{dt} = 4\text{L/min} \cdot \text{g/L} - 2\text{L/min} \cdot x(t)/(200 + 2t)\text{L}$$

Notice that, the volume of the solution in the tank increases by 2L = $(4 - 2)$L each minute; hence, after t minutes, we will have $200 + 2t$ L of solution in the tank.

We have to solve the initial value problem:

$$\begin{cases} x' = 4 - \frac{2x(t)}{(200+2t)} \\ x(0) = 0 \end{cases} \tag{5.8}$$

The integrating factor is given by:

$$u(t) = e^{\int \frac{2}{200+2t} dt} = 100 + t$$

We proceed applying the formula for the solution (5.5):

$$x(t) = \frac{c}{t + 100} + \frac{2t^2}{t + 100} + \frac{400t}{t + 100}$$

If we impose the condition $x(0) = 0$, we can determine the constant c, which is equal to zero:

$$x(t) = \frac{2t(t + 200)}{t + 100}$$

We can, then, easily determine the salt present in the tank after two and a half hours:

$$x(150) = 420\,\text{g}$$

We conclude this section with a more complicated mixing problem, where we have two tanks and consequently a system of differential equations.

In general, systems of differential equations can be effectively solved by computing eigenvalues and eigenvectors of the matrix associated with the system. However, since this study is beyond the scope of our discussion, we will limit ourselves only to particular cases, where we do not need such general theory. For more details, we refer the reader to [2].

Example 5.7.3 Suppose we have two tanks containing, respectively, 100 and 300 L of solution. In the first tank, we have, initially, 2 kg of salt, while in the second tank, we have 1 kg of salt. In the first tank, pure water enters at the rate of 4 L/s. At the bottom of the first tank, we have a drain, leading to the second tank, where the solution exits at the rate of 4 L/s. Finally, the water flows out of the second tank, at the rate of 4 L/s, so that the amount of solution in each tank remains constant. We want to determine the quantity of salt in each tank as time varies.

We proceed in a similar way as we did previously. We denote with $x(t)$ and $y(t)$ the quantity of salt in the first and second tank, at time t.

We set up two differential equations, one for each tank, and we put them into a system, as we want to find both $x(t)$ and $y(t)$.

$$
\begin{cases}
x'(t) = -\frac{1}{25}x(t) \\
y'(t) = \frac{1}{25}x(t) - \frac{1}{75}y(t)
\end{cases}
$$

The solution of the first equation is given by: $x(t) = 2e^{-t/25}$, since initially we have 2 kg of salt (this is a Malthusian law).

We substitute in the second equation:

$$
y'(t) = \frac{2}{25}e^{-t/25} - \frac{1}{75}y(t)
$$

Solving through the formula (5.5) and imposing the initial condition $y(0) = 1$, we get:

$$
y(t) = e^{-t/25}(4e^{2t/75} - 3)
$$

5.8 MALTHUSIAN LAWS AND POPULATION DYNAMICS

We want to describe an application of linear differential equations to model effectively population growth and decline.

We introduce a generalization of the Malthusian law, taking into account a constant term of growth or decline $f(t)$:

$$\frac{dN}{dt} = a(t)N(t) + f(t) \tag{5.9}$$

As usual, $a(t)$ is the population growth rate, but notice that, in this modified version of Malthusian law, it is not constant, but depends on time. In this way, we are able to model effectively a situation where the population, counting $N(t)$ individuals, besides growing with a rate proportional to its numerosity, is also subject to factors that dampen its rate of variation $N'(t)$, when $f(t) < 0$, or enhance it, in the case of $f(t) > 0$. Notice also that, unlike the usual Malthusian law examined in the previous chapters, we have that, in the formula (5.9), the coefficient $a(t)$ can vary over time.

Let us see an example.

Example 5.8.1 In 1960, the population of Guatemala, a South American country, counted about 4.1 million people, with a rate of change (births net of deaths), proportional to the existing population, amounting to 14% a year. Around 1960, the population of Guatemala decreased about half a million people per year, due to emigration. We want to determine the size of this population as time varies and predict the number of people living in Guatemala in 1965 and in 1970.

This is a Malthusian model (5.9), where the rate of annual change is constant, $a = 0.14$, and we have, due to emigration, $f = -0.5$ million a year. Let us set up the initial value problem, using the equation (5.9):

$$\begin{cases} \frac{dN}{dt} = 0.14N(t) - 0.5 \\ N(0) = 4.1 \end{cases}$$

Through the formula (5.5) of Section 5.6, we immediately have the solution of the linear differential equation, imposing the initial condition $N(0) = 4.1$:

$$N(t) = 3.571 + 0.529e^{0.14t}$$

Now, let us compute the size of the population in 1965 and in 1970. These years correspond to 5 and 10 years, respectively, after we start our observation:

$$N(5) \cong 4.6, \qquad N(10) \cong 5.7$$

From Wikipedia, we see that the recorded data are: 4.7, 5.4. This is not too far from our forecast, though it must be noted that variations in the size of an actual population depend on many more parameters, that we did not account for.

We now want to examine a small variation of this example.

Example 5.8.2 Let us look at the previous example regarding Guatemala population. We see what happens in our population modelling, if the rate of change (births net of deaths) is 7% per year, instead of 14% per year, while the emigration is the same, that is half a million people per year.

We note that the differential equation has mathematically the same form:

$$\begin{cases} \frac{dN}{dt} = 0.07N(t) - 0.5 \\ N(0) = 4.1 \end{cases}$$

We get:

$$N(t) = 7.142 - 3.042e^{0.07t}$$

However, we see that a minus sign appears in the solution: this corresponds to the fact that, due to emigration, it is theoretically possible that the population will die out.

Let us see what would have happened in 1970:

$$N(10) = 1.014$$

This is a lower value than the one we computed previously: with this model, the population in 1970 is just one million people. We can also compute when the population would be extinct, due to emigration:

$$N(t_{est}) = 0 = 7.142 - 3.042e^{0.07t_{est}} \implies t_{est} = 12.2$$

So, with these parameters, the population would become extinct after about 12 years. It is clear that many socio-economic factors contribute to the extinction of a population, so this is just an estimate. However, Malthusian equations can effectively help us to predict the trends of populations, quite accurately in the short term.

5.9 HOMOGENEOUS EQUATIONS

Homogeneous equations are first-order differential equations of the form:

$$y'(x) = \frac{P(x,y)}{Q(x,y)}$$

where $P(x,y)$ and $Q(x,y)$ are homogeneous polynomials of the same degree. We recall that a polynomial $P(x,y)$ is homogeneous of degree n if $P(tx,ty) = t^n P(x,y)$. For example, the polynomial $x^2 - 2xy$ is homogeneous of degree 2, while the polynomial $x^2 + y^2 - 2$ is not.

To solve these differential equations, we impose:

$$y = v(x)x \quad \Longrightarrow \quad \frac{dy}{dx} = \frac{dv}{dx}x + v$$

Once we have performed this substitution, we always have that the equation thus obtained is a separable equation.

Example 5.9.1 We want to solve the differential equation:

$$x + y = 2xy'$$

If we write it in normal form, we see that it is an homogeneous equation:

$$y' = \frac{x+y}{2x}, \qquad x \neq 0$$

We substitute $y = vx$, $y' = v'x + v$:

$$v'x + v = \frac{1}{2} + \frac{v}{2}$$

So

$$v' = \frac{1}{2x}(1 - v)$$

It is a separable equation. We proceed to the solution by separating the variables:

$$\int \frac{1}{v-1}dv = -(1/2)\int \frac{1}{x}dx \quad \Longrightarrow \quad \log|v-1| = -(1/2)\log|x| + c$$

So, (for $v - 1 > 0$ and $x > 0$):

$$v - 1 = k\frac{1}{\sqrt{x}} \quad \Longrightarrow \quad \frac{y}{x} = k\frac{1}{\sqrt{x}} + 1$$

Hence, we have:

$$y(x) = k\sqrt{x} + x$$

We leave the discussion for $v - 1 < 0$ to the reader.

Sometimes we cannot reach an explicit solution, but we arrive only to a solution in *implicit form*, that is, we cannot obtain y explicitly as a function of x. Let us see an example of this situation.

Example 5.9.2 We want to solve $y'(x) = \frac{x+y}{y}$. We substitute $v = y/x$ ($x \neq 0$, $y \neq 0$):

$$v'x + v = \frac{1+v}{v}$$

From which:

$$v'(x) = \frac{1}{x}\left[\frac{1+v}{v} - v\right]$$

We solve:

$$\int \frac{vdv}{1+v-v^2} = \int \frac{1}{x}\,dx$$

Computing the integrals, we get:

$$(1/10)(-(5+\sqrt{5})\log(1+\sqrt{5}-2v)+(-5+\sqrt{5})\log(-1+\sqrt{5}+2v)) = \log|x|+c$$

It is clear that, substituting $v = y/x$, we are unable to obtain y as a function of x. However, we have a solution in implicit form, since in this equality the derivative $y'(x)$ does not appear.

5.10 AUTONOMOUS DIFFERENTIAL EQUATIONS

We say that a first-order differential equation in normal form is *autonomous* if:

$$y'(t) = f(y(t))$$

that is, the function f does not explicitly depend on t. For example, the differential equation $y'(t) = 2y(t)$ is autonomous, while $y'(t) = 2t$ is not. Many of the linear equations we have studied, as Malthusian laws, are autonomous. For these equations, we can study what occurs at *equilibrium points*, that is, understand whether or not the solution tends to some constant value, when t approaches infinity, without actually solving the equation.

Definition 5.10.1 Let $y'(t) = f(y(t))$ be an autonomous differential equation. We define *equilibrium points* the values of y such that $f(y(t)) = 0$. Given an equilibrium point y_0, the constant function $y(t) = y_0$ is called an *equilibrium solution*.

Let us take a look at a simple example.

Example 5.10.2 Let us consider the autonomous differential equation: $y'(t) = -2y(t)$ and compute the equilibrium points:

$$y' = -2y = 0 \quad \Longrightarrow \quad y = 0$$

Hence, the equilibrium solution is $y = 0$. If we draw the field of directions, we immediately see that all solutions tend to the equilibrium solution:

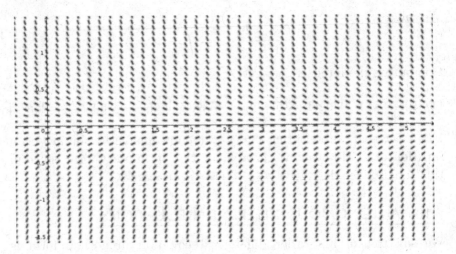

Now let us see what happens, if we consider an equation very similar to $y' = -2y$, having the same equilibrium point, but with a very different behavior at equilibrium:

$$y' = 2y = 0 \quad \Longrightarrow \quad y = 0$$

In this case, if we draw the field of directions, we see that the solutions quickly depart from the equilibrium solution $y = 0$, when t goes to infinity.

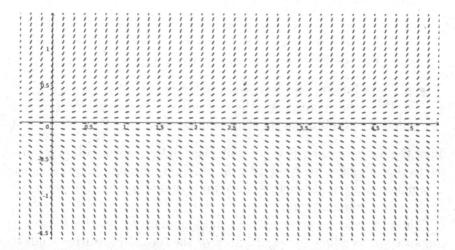

We want to characterize the behavior of equilibrium solutions of an autonomous differential equation and understand when one of the two quite different situations described above occurs.

The previous example suggests the following definition.

We give an intuitive definition of the concept of stability: it is possible to give a more rigorous definition, using the terminology we introduced when we studied limits, but this discussion goes beyond the purposes of this elementary textbook.

Definition 5.10.3 Let us consider the autonomous differential equation $y'(t) = f(y(t))$.

1. We say that an equilibrium point y_0 is *asymptotically stable*, if there exists an interval (a, b), containing y_0, such that every solution of the equation, with initial condition $y(t_0)$ in (a, b), approaches the equilibrium solution $y(t) = y_0$, as t tends to infinity.

2. We say that an equilibrium point y_0 is *unstable*, if every solution of the equation, with initial condition in an interval containing y_0, moves away from the equilibrium solution $y(t) = y_0$, as t tends to infinity.

Examining the direction field of the equation $y' = -2y$, we see that $y = 0$ is an asymptotically stable equilibrium point, while for the equation $y' = 2y$, the point of equilibrium $y = 0$ is unstable. Actually, looking at the examples seen so far, given an equilibrium point y_0, we have that:

- if the solutions are increasing (their derivative is positive), for $y < y_0$ and decreasing (their derivative is negative), for $y > y_0$, then the equilibrium point y_0 is asymptotically stable;

- if the solutions are decreasing, (the derivative is negative), for $y < y_0$ and increasing, (the derivative is positive), for $y > y_0$, then the equilibrium point y_0 is unstable.

Since the derivative of $y(t)$ is given by $f(y(t))$, the previous observations tell us that, to have an asymptotically stable equilibrium point, y_0 must be a maximum point, while, to have an unstable equilibrium point, y_0 must be a minimum point.

These simple observations can be summarized by the following proposition, recalling the significance of the second derivative studied in Chapter 3.

Proposition 5.10.4 *Let $y'(t) = f(y(t))$ be an autonomous differential equation, y_0 an equilibrium point. Then, we have that:*

- *If $f'(y_0) < 0$, then y_0 is an asymptotically stable equilibrium point.*

- *If $f'(y_0) > 0$, then y_0 is an unstable equilibrium point.*

Note that, if $f'(y_0) = 0$, then we cannot say anything on the equilibrium point y_0.

Let us see an example.

Example 5.10.5 We want to study the equilibrium points and their stability, for the autonomous differential equation:

$$y' = y^2 - 1$$

The equilibrium points are $y = \pm 1$. Analyzing the sign of $y^2 - 1$, we immediately see that $y = -1$ is an equilibrium point, which is asymptotically stable, while $y = 1$ is an unstable equilibrium point.

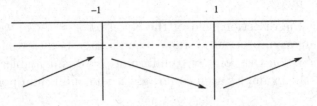

If we want to apply the previous proposition and confirm our reasoning we see that, since $f'(y) = 2y$:

- $f'(-1) = -2 < 0$, so $y = -1$ is an asymptotically stable equilibrium point;

- $f'(1) = 2 > 0$, so $y = 1$ is an unstable equilibrium point.

The field of directions shows us that this reasoning is correct:

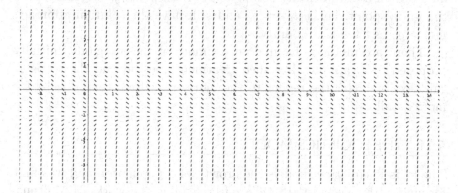

Indeed, the solutions with initial value $y(t_0)$, in a suitable interval containing $y = 1$, will depart from this value as time increases, while the solutions $y(t)$, with initial value in a (small) interval containing $y = -1$, will approach this value for very large t. In other words, as we observe from the graph, we have that $\lim_{t\to\infty} y(t) = -1$. Notice that we cannot choose arbitrarily the initial condition, if we want the solution to approach $y = -1$ for large t. For example, if we choose an initial value $y(t_0) = 2$, then the solution will approach infinity, as t tends to infinity.

To graphically represent these behaviors, we draw the *phase line*, represented here by the y axis:

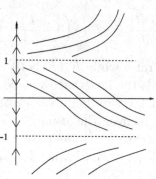

The arrows, on the y axis, express whether the equilibrium point is asymptotically stable or unstable.

We conclude this section with some observations. We can extend these reasonings to the case of systems of differential equations: we will have in this case not the phase line, but the phase space. It is especially interesting to understand the concept of *basin of stability*, that is, how far from an asymptotically stable equilibrium point y_0 we can choose the initial point, i.e., $y(t_0) = y_0$ and still have the solution approach $y(t) = y_0$ as t tends to infinity.

There are methods to give an estimate of these regions, initially proposed by Lyapunov, but such treatment goes beyond the scope of this text; we refer the reader to the more specialized textbook [2] for more details.

5.11   THE LOGISTICS MODEL

The logistic equation models the growth of a population and was first formulated by Verhulst. In 1838, after reading the works of Malthus, Verhulst built a model to describe the time evolution of a population. Verhulst set the population growth rate as proportional to the size of the population, as it is the case for Malthusian laws, but with a correction term to account for the available resources, generally limited by the environment. Hence, it models very effectively situations in which the growth of a population is limited in some way: for example, if we have a population of bacteria, which grows in a laboratory, with limited resources, e.g., the nutrients in the Petri dish.

Let us now discuss mathematically the logistic equation. Suppose that $P(t)$ represents the size of a population at time t. The logistic differential equation is given by:

$$\frac{dP(t)}{dt} = rP(t)\left(1 - \frac{P(t)}{K}\right), \tag{5.10}$$

where r is the growth rate, while K $(K > 0)$ is the *carrying capacity* (or *capacity*) and models the fact that we have limited resources.

Indeed, we see that for very large values of K, we can ignore the term $P(t)/K$ and we recover the usual Malthusian equation. This corresponds to the fact that, if we have abundant resources, the population growth follows a Malthusian law, i.e., $P(t)$ increases (or decreases depending on the sign of r) exponentially. In the following observation, we put into practice what we have learnt in the previous section.

Observation 5.11.1 The logistic equation (5.10) is an autonomous equation and we can study equilibrium solutions and their stability, without actually solving the equation. Let us compute the equilibrium points:

$$rP(t)\left(1 - \frac{P(t)}{K}\right) = 0 \qquad \Longrightarrow \qquad P = 0,\, P = K$$

Studying the sign, we immediately see that $P = 0$ is an unstable solution, while $P = K$ is an asymptotically stable solution. We can draw the phase line:

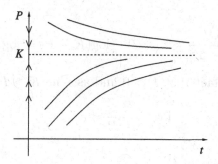

Now we interpret the above graph: as soon as the population $P(t)$ assumes positive, but small, values, then $P(t)$ approaches the value K, i.e., the capacity:

$$\lim_{t \to \infty} P(t) = K$$

The solution $P = K$ is asymptotically stable; the size of the population tends to the capacity K, regardless of the initial condition. On the other hand, the solution $P = 0$ is unstable. From a mathematical point of view, we have that the initial condition $P(t_0)$ can be either greater or smaller than K (we do not consider the case of negative population). However, since we are interested in applications, we always assume that $P(t_0) < K$, because in our model, due to limited resources, $P(t)$ cannot exceed the capacity K. Hence, the interval in which we are interested in studying $P(t)$ is $(0, K)$. When we solve the differential equation (5.10), we will see more explicitly the interval of existence of the solution.

5.12   SOLUTION OF THE LOGISTIC EQUATION

In this section, we want to solve the logistic equation:

$$\frac{dP(t)}{dt} = rP(t)\left(1 - \frac{P(t)}{K}\right) \qquad (5.11)$$

Since r and K are constant, it is a separable equation:

$$\int \frac{K dP}{P(K - P)} = \int r\, dt \qquad (5.12)$$

We use the method of partial fractions described in the previous chapter:

$$\frac{K}{P(K - P)} = \frac{1}{P} + \frac{1}{K - P}$$

Hence:

$$\int \frac{K dP}{P(K - P)} = \log(P) - \log(K - P) = \log \frac{P}{K - P}$$

where we use the fact that $P > 0$ and assume $K > P$. Hence:

$$\log \frac{P}{K - P} = rt + c$$

From which:

$$\frac{P}{K - P} = Ce^{rt}, \qquad C = e^{c}$$

Suppose $P(t_0) = P_0$. If we substitute:

$$Ce^{rt_0} = \frac{P_0}{K - P_0} \qquad \Longrightarrow \qquad C = \frac{P_0 e^{-rt_0}}{K - P_0}$$

For simplicity, we assume $t_0 = 0$ (the general case being a small variation). We obtain:

$$P(t) = \frac{KP_0 e^{rt}}{K + P_0\left(e^{rt} - 1\right)}$$

We note that, during the calculation, we impose that $0 < P(t) < K$. Hence, for $0 < P(t_0) < K$, the only interesting case for applications, we have that the interval of existence of the solution is $0 < P(t) < K$.

Let us see an example of application to a concrete problem.

Example **5.12.1** We consider a small urban forest, that can support, at most, a population of 1000 squirrels. If initially we have 100 squirrels, that reproduce at an annual rate of 0.4, we want to compute after how many years we will have 400 squirrels.

We model the time evolution of the population $P(t)$ of squirrels, through the logistic equation:

$$P'(t) = 0.4P(t)\left(1 - \frac{P(t)}{1000}\right)$$

where we set $r = 0.4$ and $K = 1000$, which represents the capacity.

We proceed to the solution as we described above. The initial condition, for $t = 0$, gives us immediately:

$$C = \frac{P_0}{K - P_0} = \frac{100}{1000 - 100} = 0.9$$

Substituting into the formula we get:

$$P(t) = \frac{1000}{1 + 9e^{-0.4t}}$$

Let us now compute how many years are necessary to reach a population of 400 squirrels:

$$\frac{1000}{1 + 9e^{-0.4t_f}} = 400 \quad \Longrightarrow \quad t_t = 4.48$$

Hence, we will have a population of 400 squirrels after four and a half years.

5.13 EXERCISES WITH SOLUTIONS

5.13.1 Solve the differential equation $y'(x) = \frac{e^{3x}}{y}$.
Solution. It is a separable equation. Hence:

$$\int y\,dy = \int e^{3x}\,dx \quad \Longrightarrow \quad y^2 = (2/3)e^{3x} + c$$

So we have $y(x) = \pm\sqrt{(2/3)e^{3x} + c}$, i.e., we get two possible solutions, the sign of the square root is determined, once we have an initial condition.

5.13.2 Solve the initial value problem $y'(x) = \frac{e^{3x}}{y}$, $y(0) = -2$.

Solution. From the previous exercise, we have $y(x) = \pm\sqrt{(2/3)e^{3x} + c}$. Since $y(0) = -2$, we take the negative sign in the expression of the solution:

$$-\sqrt{(2/3)e^0 + c} = -2 \implies 2/3 + c = 4 \implies c = 10/3$$

Hence, the solution to the initial value problem is given by:

$$y(x) = -\sqrt{(2/3)e^{3x} + 10/3}$$

5.13.3 Solve the differential equation $y'(x) = \frac{2y+x}{3x}$.

Solution. The given equation is in normal form and it is both a linear and an homogeneous equation. Hence, we can proceed to the solution by choosing one of the two methods described above for solving these equations. We choose the formula for the linear equation, $u(x) = x^{2/3}$, obtaining $y = cx^{2/3} + x$.

5.13.4 Solve the Cauchy problem:

$$\begin{cases} y'(x) = \frac{2y+x}{3x} \\ y(1) = 2 \end{cases}$$

Solution. We have to just impose the initial condition to the solution we found in the previous exercise. We obtain $c = 1$; hence, the solution of the Cauchy problem is given by $y = x^{2/3} + x$.

5.13.5 A cup of coffee is served at a temperature 95^oC and left on the counter. The cafeteria is air-conditioned with an ambient temperature of 20^oC. After 5 minutes, the temperature of the coffee is 45^oC. Determine how long it takes for the coffee to cool down at a temperature of 22^oC.

Solution. According to Newton's law, we have:

$$\frac{dT}{dt} = -k(T - T_e)$$

Let us separate the variables:

$$\int \frac{dT}{T - T_e} = \int -k\,dt + c$$

or

$$\log(T - T_e) = -kt + c$$
$$T - T_e = e^{-kt} c'$$
$$T = T_e + c'e^{-kt}$$

From the initial condition $T(0) = 95°C$ we find, evaluating in $t = 0$:

$$95°C = 20°C + c' \quad \implies \quad c' = 75°C$$

From the given condition $T(5\text{min}) = 45°C$ we find, evaluating in $t = 5\text{min}$:

$$45°C = 20°C + 75C° e^{-k\,5\text{min}} \quad \implies \quad k = \frac{\log 3}{5} \text{min}^{-1}$$

So the equation for the temperature is:

$$T = 20°C + 75°C e^{-(\frac{\log 3}{5} \text{min}^{-1})t}$$

Now, we can find after how many minutes the coffee cools down to $22°C$:

$$22°C = 20°C + 75°C e^{-(\frac{\log 3}{5} \text{min}^{-1})t} \quad \implies \quad t = \frac{5\log(75/2)}{\log(3)} \sim 16\text{min}$$

5.13.6 In a reservoir serving a small town, there are, initially, 100 million L of water containing a total of 5 kg of fluoride. To reduce the fluoride content, pure water enters the reservoir at the rate of 3 million L/day, while an equal amount of water exits at the same rate.

a. Determine, through a differential equation, the quantity of fluoride $x(t)$ in the reservoir at time t.

b. Determine how many days are necessary to have a fluoride concentration equal to 1.5 μ/L.

c. Assume now that we start pumping 3 million L daily into a second reservoir containing initially 100 million L of pure water. At the same time, water is flowing out from the second reservoir, so that its total volume is constant (100 million L). Compute the amount of fluoride $y(t)$ in the second reservoir and its value for t approaching infinity.

Solution. a. We first have to solve the differential equation with initial condition:

$$x'(t) = -3\frac{x(t)}{100}, \qquad x(0) = 5$$

We immediately have $x(t) = 5e^{-0.03\,t}$.

b. To determine the time t_f, we impose:

$$\frac{5\mathrm{kg}\,e^{-0.03\,t_f}}{100 \times 10^6 \mathrm{L}} = 1.5 \times 10^{-6}\,\mathrm{g/L}$$

Hence, $t_f \cong 117$ days.

c. We have the linear differential equation with initial condition:

$$y'(t) = \frac{15}{100}e^{-0.03\,t} - (3/100)y(t), \qquad y(0) = 0$$

We get $y(t) = \frac{15}{100}e^{-0.03t}t$. For t approaching infinity, we have that $y(t)$ tends to zero.

5.13.7 The isotope of Uranium-234 and the isotope of Thorium-230 are radioactive. Uranium-234 decays into Thorium-230 with a half-life equal to $t_{1/2}^U = 245,000$ years, while Thorium has a half-life $t_{1/2}^T = 75,380$ years. If we have initially 100 g of Uranium, determine how much Thorium we will have after $200,000$ years.

Solution. Let us focus first on uranium, which follows a law of exponential decay (see Chapter 1). We denote by $U(t)$ the quantity of uranium at time t, expressed in thousands of years, and with U_0 the initial amount of uranium. We have that:

$$U(t) = U_0 e^{-\lambda_U t} \tag{5.13}$$

To determine the constant λ_U, we use the information given by the half-life, i.e., the time required for the uranium to reduce to half of its initial amount. We can express this condition as follows:

$$\frac{U_0}{2} = U_0 e^{-\lambda_U t_{1/2}^U} \tag{5.14}$$

which brings us to

$$e^{-\lambda_U t_{1/2}^U} = \frac{1}{2} \quad \Longrightarrow \quad \lambda_U = \frac{\ln 2}{t_{1/2}^U} = 0.0028 \tag{5.15}$$

where we express λ_U in 10^{-3} years^{-1}. So, the expression for the amount of uranium at time t is given by

$$U(t) = 100e^{-0.0028t} \tag{5.16}$$

Now, by analogy with the mixing problem we solved previously, let us think about the variation of Uranium-238 and Thorium-230: Uranium-234 is transformed into Thorium-230, and Thorium-230 decays into other isotopes. If we denote by $U(t)$ e $T(t)$ their respective quantities, we obtain the system:

$$\begin{cases} U'(t) = -\lambda_U U(t) \\ T'(t) = \lambda_U U(t) - \lambda_T T(t) \end{cases}$$

where λ_T is the decay constant of Thorium, which we can compute similarly to λ_U:

$$\lambda_T = \frac{\ln 2}{t_{1/2}^T} = 0.0092$$

We substitute $U(t)$:

$$T'(t) = \lambda_U(100e^{-0.0028t}) - \lambda_T T(t)$$

It is a linear equation that we can solve with the formula (5.5). The amount of Thorium-230 at time t, taking into account that $T(0) = 0$, is given by

$$T(t) = 43.75[e^{-0.0028t} - e^{-0.0092t}] \tag{5.17}$$

After $100,000$ years (so $t = 100$) we have

$$T(100) \simeq 15.6\text{g} \tag{5.18}$$

5.13.8 A patient receives intravenously a drug at a rate of 20 mg/h. The drug is metabolized by the organism as follows: every two hours the quantity of drug in the blood is cut in half. Compute the quantity of drug in the blood as time varies and establish, if possible, after how many hours the drug concentration exceeds 10 mg/L.

Solution. We denote by $x(t)$ the amount of drug expressed in milligrams. Since we have half-life $t_{1/2} = 2$ hours, we can immediately compute the constant λ, which allows us to express the amount of drug $x(t)$ *without* infusion:

$$x'(t) = \lambda x(t), \qquad \lambda = \log(2)/t_{1/2} = 0.35\text{h}^{-1}$$

We now proceed to set up the differential equation which takes into account the infusion:

$$x'(t) = -0.35x(t) + 20\text{mg/h}$$

It is a linear equation. By imposing the condition $x(0) = 0$, we obtain the solution:

$$x(t) = 57.14 - 57.14e^{-0.35t}$$

Now let us see if the concentration of 10 mg/L is reached after a time t_f:

$$x(t_f) = 57.14 - 57.14e^{-0.35t_f} = 10 \times 5$$

where we assume that (on average) there are 5 L of blood per every individual. We see with a calculation that $t_f \cong 6$ hours.

5.13.9 An epidemic breaks out in a village of 1000 people. If initially there are 80 infected people and after four weeks we have 500 infected people, use a logistic model to predict after how many weeks we will have 90% of the village people infected by the virus.

Solution. The logistic equation is given by:

$$\frac{dP(t)}{dt} = rP(t)\left(1 - \frac{P(t)}{K}\right)$$

where capacity $K = 1000$ and we need to determine r. From the formula:

$$P(t) = \frac{KP_0e^{rt}}{K+P_0(e^{rt}-1)} = \frac{1000 \times 80e^{rt}}{1000+80(e^{rt}-1)}.$$

Since after 4 weeks we have 500 infected people, we get:

$$P(4) = \frac{80,000e^{4r}}{1000 + 80\left(e^{4r} - 1\right)} = 500$$

From which we obtain $r = 0.61$. We now proceed to compute t_f the time, expressed in weeks, needed to reach 900 infected people:

$$900 = \frac{1000 \times 80e^{0.61t_f}}{1000 + 80\left(e^{0.61t_f} - 1\right)}$$

From which $t_f = 7.6$. Hence, in about 7 weeks and a half, we will have 90% of the people in the village infected with the virus.

5.14 SUGGESTED EXERCISES

5.14.1 a. Draw the field of directions of the following differential equations:

1. $xy' = xy + 3$

2. $y' = \sin(x) - y^2$

3. $x^2 + y' = x - 2$

4. $xy = y'$

5. $y' = \log(x) + x^2$

6. $y' = xe^y$

b. For each equation, state whether it is separable and, if so, solve it.

5.14.2 Establish the correspondence between the direction fields and the following differential equations:

$$1.\, y' - 2y = 3e^x, \qquad 2.\, y' = y^2 - x$$

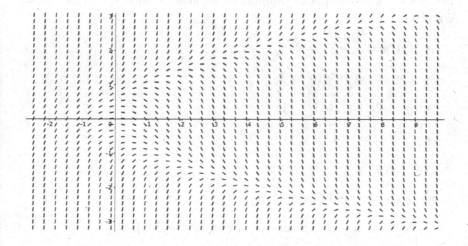

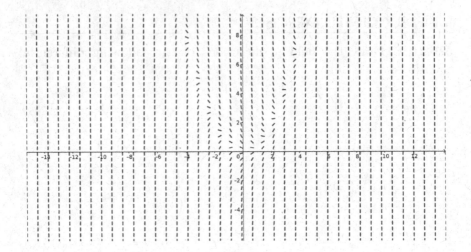

5.14.3

Solve the following linear differential equations:

- $y' + y = 3$

- $y' + 2ty = 4t$

- $ty' = 2t + t^4$

- $xy' + y = \sin(x)$

5.14.4

Solve the following linear differential equations with the given initial condition:

- $y' + y = 3$, $y(0) = 2$.

- $y' + 2ty = 4t$, $y(0) = 1$.

- $ty' = 2t + t^4$, $y(1) = 1$.

- $xy' + y = \sin(x)$, $y(1) = -1$.

5.14.5 Let $y' = \frac{y+x}{x-1}$ be given.

a. Draw the field of directions

b. Say if it is (more than one option is possible):

1. linear

2. separable

3. homogeneous

4. in normal form

5. autonomous

   c. Solve the given equation by two different methods, with the initial condition $y(2) = 1$.

5.14.6 Let us consider a population, whose numerosity at time t, expressed in years, we denote by $N(t)$. Assume the population reproduces at a rate one percent growth per year. However, every year we have an emigration of 1 million people. If $N(0) = 50$ million, is there a time in which the population becomes extinct?

5.14.7 A tank with capacity of 50 L holds 20 L of pure water at time $t = 0$. A 0.5 g/L solution of salt enters at a rate of 4 L/min. At the same time, the water exits at a rate of 2 L/min. How much salt is in the tank when it is completely filled?

5.14.8 A lake of constant volume contains 50 km^3 of water and receives from its tributaries an amount of 24 km^3/year. Some factories pour contaminated water into the lake at the rate of 1 km^3/year. If the percentage of contaminated water exceeds 0.8% of the total, the lake becomes unsuitable for aquatic life. What should the input of contaminated water be, so that 0.8

5.14.9 A patient receives a drug intravenously at a rate of 10 mg/h. An hour later, the concentration of the drug in the patient is of 1 mg/L. Assuming the patient has 5 L of blood and the drug is metabolized at a rate proportional to the amount of drug in the body, find the drug elimination rate (assuming it is constant). Finally, determine the long-term concentration of the drug in the patient's body.

5.14.10 A glucose solution is given intravenously to a patient at constant velocity v. As the glucose enters the bloodstream, it is transformed in other substances and eliminated with a speed proportional to its concentration $C(t)$, where the proportionality constant is $k = 5/$minutes.

a. Assuming that at time $t = 0$ the patient is in an hypoglycemic condition, with $C(0) = 60$ mg/dL and $v = 350$ mg/dL/minute (dL=deciliter), determine the concentration $C(t)$ as time varies.

b. Determine the time (if it exists) for which the concentration is 70 mg/dL (minimum concentration of a healthy subject)?

c. What speed v of administration allows to have a concentration equal to 80 mg/dL for t approaching infinity? Is this equilibrium solution stable?

5.14.11 Bertalanffy's model of tumor growth takes into account both cell reproduction and death. Suppose the growth of the tumor is proportional to the surface of the tumor, as the nutrient enters through the surface and the cell death rate is proportional to the volume of the tumor. The Bertalanffy's model is given by:

$$V'(t) = aV^{2/3} - bV$$

where a and b are two positive constants and depend from the type of cancer and other conditions.

1. Study the equilibrium points and their stability.
2. Determine the solution with $a = b = 1$.

5.14.12 Solve the homogeneous equations:

- $x^2 + y^2 = 2xyy'$

- $x + y + (y - x)y' = 0$

5.14.13 Given the following differential equations say if they are separable, linear, homogeneous and solve them.

1. $xy' + x^2 = y$.

2. $y^2 = yy' + 3x$. (Help: set $u = y^2$).

3. $x - yy' + e^{y^2 - x^2} = 0$. (Help: set $u = x^2 - y^2$).

5.14.14 a) For each of the following autonomous equations, determine the equilibrium points and say if they are asymptotically stable or unstable.

1. $y' = y^3 - 3y$

2. $y' = y^2 - y - 1$

3. $x' = x^4 - 1$

4. $x' = 2x - 9$

b. For each equation, draw the phase line.

5.14.15 The following autonomous differential equations represent the function giving a population size (expressed in millions of people) in terms of the time t. Determine the stability of the equilibrium points and what happens to the population size if it assumes a value near an equilibrium point.

1. $N'(t) = 2 - 3N$

2. $N'(t) = 2N - 3N^2$

3. $N'(t) = 3N^2 - 2N$

4. $N'(t) = N^2 - 3N + 2$

5.14.16 Suppose that, in an isolated village of 10,000 inhabitants there are initially 10 people infected with a virus. After 5 weeks, we have that 10% of the people in the village are infected. Use the logistic equation to determine when 99% of the people in the village will be infected.

5.14.17 In a mountain pond we release initially ten trouts. After six months, we estimate that there are about 100 trouts, while after two years the estimated number raises to 700. Use the logistics model to establish the capacity of the pond.

[Note: in the first 6 months, we can use as approximation a Malthusian model, as the capacity is much greater than the population.]

CHAPTER 6

Second-Order Differential Equations

6.1 CAUCHY'S THEOREM

We want to solve the following problem: determine the solution of a second-order differential equation together with initial conditions. This problem takes the form:

$$\begin{cases} y'' + p(t)y' + q(t)y = g(t) \\ y(t_0) = y_0 \\ y'(t_0) = y_0'. \end{cases} \tag{6.1}$$

This problem is called the *initial value* or *Cauchy's problem* for second-order differential equations. We have the existence and uniqueness theorem, very similar to the one stated in the case of first-order differential equations (see Chapter 5 for a comparison).

Theorem 6.1.1 *Let $p, q, g : (a, b) \longrightarrow \mathbb{R}$ be continuous functions and $t_0 \in (a, b)$. Then, there exists a unique solution $y : (a, b) \longrightarrow \mathbb{R}$ of the Cauchy problem:*

$$\begin{cases} y''(t) + p(t)y'(t) + q(t)y(t) = g(t) \\ y(t_0) = y_0 \\ y'(t_0) = y_1 \end{cases} \tag{6.2}$$

DOI: 10.1201/9781003343288-6

The proof of this result is too technical for the present treatment, we refer the interested reader to [3] for more details.

Let us see an example of a very simple application of this powerful result to understand better the role of initial conditions.

Example 6.1.2 We want to solve the differential equation:

$$y''(t) = 2$$

We need to find a function whose second derivative is constant. Let us proceed as follows:

$$y'(t) = 2t + c_1 \quad \Longrightarrow \quad y(t) = t^2 + c_1 t + c_2$$

We have found the *general solution* of the given equation, that is, an expression for $y(t)$ that comprehends all solutions, once we set values of the constants c_1 and c_2.

We now turn to the solution of the following Cauchy problem:

$$\begin{cases} y''(t) = 2 \\ y(1) = -1 \\ y'(1) = 3 \end{cases}$$

We impose initial conditions on the expressions of y and y':

$$y'(1) = 2 + c_1 = 3, \ y(1) = 1 + c_1 + c_2 = -1 \quad \Longrightarrow \quad c_1 = 1, c_2 = -3$$

Hence, the solution to the given Cauchy problem is:

$$y(t) = t^2 + t - 3$$

Notice that, since we integrated twice, we obtained two integration constants c_1, c_2, that we determine, imposing the two initial conditions.

Unlike the case of first-order differential equations, we will see that, in general, we will not integrate to find the solution of second-order linear equations; instead, we will make use of some particularly simple functions and then construct the general solution starting from them.

6.2 THE WRONSKIAN

The Wronskian is of fundamental importance in the theory of differential equations. However, as we will see from its definition, it is simply a function associated with a pair of differentiable functions, with no direct reference to differential equations. Hence, it is not immediately clear how the Wronskian is related with the problem of determining the general solution of a differential equation.

Definition 6.2.1 Let y_1 and y_2 be two differentiable functions in an interval $I \subset \mathbb{R}$. We define their *Wronskian* as:

$$W(t) = y_1(t)y_2'(t) - y_2(t)y_1'(t), \qquad t \in I$$

Let us look at an example.

Example 6.2.2 We compute the Wronskian of the functions $y_1 = e^t$ and $y_2 = e^{2t}$ in the interval $I = \mathbb{R}$:

$$W(t) = 2e^t e^{2t} - e^t e^{2t} = e^{3t}$$

The most interesting case is when we take the Wronskian of two solutions of the same homogeneous differential equation, that is, an equation of the type (6.1) with $g(t) = 0$.

Proposition 6.2.3 *Let y_1 and y_2 be solutions of the homogeneous differential equation:*

$$y''(t) + p(t)y'(t) + q(t)y(t) = 0$$

in the interval $I \subset \mathbb{R}$. Then, we have either $W(t) = 0$ for every t in I or $W(t) \neq 0$ for every t in I.

Proof. We calculate the derivative of the Wronskian:

$$W'(t) = (y_1(t)y_2'(t) - y_2(t)y_1'(t))' = y_1'y_2' + y_1y_2'' - y_2'y_1' - y_2y_1'' = y_1y_2'' - y_2y_1''$$

Since y_1 and y_2 are solutions of $y'' + py' + qy = 0$, we can substitute in the expression of $W'(t)$, $y_1'' = -py_1' - qy_1$ and $y_2'' = -py_2' - qy_2$:

$$W'(t) = y_1(-py_2' - qy_2) - y_2(-py_1' - qy_1) = -p(y_1y_2' - y_2y_1') = -pW(t)$$

We have obtained a separable differential equation for $W(t)$:

$$W'(t) = -pW(t) \quad \Longrightarrow \quad W(t) = W(t_0)e^{\lambda}$$

where $\lambda = -\int p(t)dt$.

Thus, we see that, if $W(t_0) = 0$, then $W(t) = 0$ for all $t \in I$. If $W(t_0) \neq 0$, then $W(t) \neq 0$ for every $t \in I$. □

Let us see an example of the case where $W(t) = 0$ for every $t \in I$.

Example 6.2.4 We compute the Wronskian of the functions $y_1 = \sin(t)$ and $y_2 = -7\sin(t)$, which are solutions of the homogeneous equation $y'' + y = 0$ in the interval $I = [0, 2\pi]$:

$$W(t) = \sin(t)(-7\cos(t)) - \cos(t)(-7\sin(t) = 0$$

However, notice that the requirement for y_1 and y_2 to be solutions of the same homogeneous equation is essential. Indeed, if we take two arbitrary functions, for example, $y_1(t) = t^2 + 1$ and $y_2(t) = t^3$, their Wronskian $W(t) = (t^2 + 1)3t^2 - 2t(t^3) = t^2(t^2 + 3)$ is equal to zero just for $t = 0$. Hence, this is an example where, neither $W(t) \neq 0$ nor $W(t) = 0$, for all t in the interval $I = \mathbb{R}$. This does not contradict the above theorem, but simply tells us that y_1 and y_2 cannot be solutions of the same homogeneous equation.

In the next section, we will use the Wronskian to determine the general solution of a homogeneous differential equation.

6.3 HOMOGENEOUS LINEAR EQUATIONS

In this section, we want to find the *general solution* of a second-order homogeneous differential equation, that is, an equation of the type:

$$y''(t) + p(t)y'(t) + q(t)y(t) = 0 \tag{6.3}$$

By the term *general solution*, we mean an expression that allows us to obtain *all* the solutions of the given equation.

Lemma 6.3.1 *Assume y_1 and y_2 are solutions of $y''(t) + p(t)y'(t) + q(t)y(t) = 0$. Then, for c_1, $c_2 \in \mathbb{R}$, we have that $c_1y_1 + c_2y_2$ is a solution of $y''(t) + p(t)y'(t) + q(t)y(t) = 0$.*

Proof. We substitute $c_1 y_1 + c_2 y_2$ into the equation $y''(t) + p(t)y'(t) + q(t)y(t) = 0$:

$$(c_1 y_1 + c_2 y_2)'' + p(t)(c_1 y_1 + c_1 y_2)' + q(t)(c_1 y_1 + c_2 y_2) =$$

$$= c_1 y_1'' + c_2 y_2'' + c_1 p(t) y_1' + c_2 p(t) y_2' + c_1 q(t) y_1 + c_2 q(t) y_2$$

$$= c_1 (y_1'' + p(t)y_1' + q(t)y_1) + c_2 (y_2'' + p(t)y_2' + q(t)y_2) =$$

$$= c_1 \cdot 0 + c_2 \cdot 0 = 0$$

$$(6.4)$$

□

Hence, we showed that, if we know two solutions y_1 and y_2 of a differential equation (6.3), all expressions of the type: $c_1 y_1 + c_2 y_2$ are also solutions. We call such an expression, a *linear combination* of y_1 and y_2 with scalars c_1 and c_2. The real numbers c_1 and c_2 are called *scalars* to distinguish them from *vectors*, which have many components and are used in physics to describe physically interesting quantities.

To solve the problem of determining all solutions of (6.3), it is necessary to recall a result on linear systems, for more details see Appendix 6.12.

Theorem 6.3.2 *Consider a linear system of two equations in two unknowns:*

$$\begin{cases} ax + by = r \\ cx + dy = s \end{cases} \qquad (6.5)$$

The system admits unique solution if and only if $ad - bc \neq 0$.

We now consider a Cauchy problem in the homogeneous case, i.e., when $g(t) = 0$:

$$\begin{cases} y'' + p(t)y' + q(t)y = 0 \\ y(t_0) = y_0 \\ y'(t_0) = y_0'. \end{cases} \qquad (6.6)$$

Suppose we know two solutions $y_1(t)$ and $y_2(t)$ of the equation $y''(t) + p(t)y'(t) + q(t)y(t) = 0$. We say that these are *particular solutions*, meaning that they do not depend on any arbitrary constant.

By Lemma 6.3.1, we know that $c_1 y_1 + c_2 y_2$ is also a solution of the given equation, for any choice of the two real constants c_1 and c_2. We impose the two initial conditions on this solution:

$$\begin{cases} c_1 y_1(t_0) + c_2 y_2(t_0) & = y_0 \\ c_1 y_1'(t_0) + c_2 y_2'(t_0) & = y_0' \end{cases}$$

We obtain a linear system where the unknowns, that is, the quantities to be determined, are c_1 and c_2. By Theorem 6.3.2, we have that the system admits a unique solution if and only if

$$W(t_0) = \det \begin{pmatrix} y_1(t_0) & y_2(t_0) \\ y_1'(t_0) & y_2'(t_0) \end{pmatrix} = y_1(t_0) y_2'(t_0) - y_2(t_0) y_1'(t_0) \neq 0$$

We can finally prove the result we are interested in, that is, how to derive the general solution of a homogeneous differential equation.

Theorem 6.3.3 *Let us consider the differential equation:*

$$y''(t) + p(t)y'(t) + q(t)y(t) = 0$$

with p, q continuous functions in an interval I. Let y_1 and y_2 be two solutions with non-zero Wronskian in $t_0 \in I$. Then, we can express any solution of the differential equation as:

$$c_1 y_1(t) + c_2 y_2(t) \tag{6.7}$$

for suitable constants c_1 and c_2.

Proof. Suppose $y(t)$ is a solution of the given equation and consider a point t_0 in the interval I. We want to prove that there are values of the constants c_1 and c_2, which allow to express $y(t)$ in the form (6.7) in the interval I.

We impose:

$$\begin{cases} c_1 y_1(t_0) + c_2 y_2(t_0) & = y(t_0) \\ c_1 y_1'(t_0) + c_2 y_2'(t_0) & = y'(t_0) \end{cases}$$

since the Wronskian of y_1 and y_2 is non-zero in the interval, by Theorem 6.3.2, we get a unique solution for $c_1 = A$ and $c_2 = B$, for suitable A and B. Hence, both $y(t)$ and $Ay_1(t) + By_2(t)$ are solutions of the same Cauchy problem. By uniqueness (Cauchy's Theorem), we have that $y(t) = Ay_1(t) + By_2(t)$. □

Let us see an example.

Example 6.3.4 We want to determine the general solution of the homogeneous equation $y''(t) - 5y'(t) = 0$, knowing that $y_1(t) = 5$ and $y_2(t) = e^{5t}$ are solutions. First we compute the Wronskian of y_1 and y_2:

$$W(t) = \det \begin{pmatrix} y_1(t) & y_2(t) \\ y_1'(t) & y_2'(t) \end{pmatrix} = y_1(t)y_2'(t) - y_2(t)y_1'(t) = 25e^{5t} \neq 0$$

Since $W(t) \neq 0$ for all real t, we can proceed and write immediately, by the previous theorem, the general solution:

$$y(t) = 5c_1 + c_2 e^{5t}$$

As we shall see later on, it is not difficult to compute two particular solutions of a given equation. Hence, we can write the general solution, using the result stated in the previous theorem and based on the concept of Wronskian.

6.4 LINEAR EQUATIONS

In this section, we want to determine the general solution of the differential equation:

$$y''(t) + p(t)y'(t) + q(t)y(t) = g(t) \tag{6.8}$$

We start with a lemma which reduces the problem to the *homogeneous* case, that is, the case in which the function $g(t) = 0$.

Lemma 6.4.1 *Let y_s be a solution of the equation*

$$y''(t) + p(t)y'(t) + q(t)y(t) = g(t)$$

Then, $y_s = y_h + y_p$, where y_p is a particular solution and y_h is a solution of $y''(t) + p(t)y'(t) + q(t)y(t) = 0$.

Proof. Let y_s be a solution. We first check that $y_h = y_s - y_p$ is a solution of $y''(t) + p(t)y'(t) + q(t)y(t) = 0$. Indeed:

$$(y_s - y_p)'' + p(t)(y_s - y_p)' + q(t)(y_s - y_p) =$$

$$= y_s'' - y_p'' + p(t)(y_s' - y_p') + q(t)(y_s - y_p) =$$

$$= y_s'' + p(t)y_s' + q(t)y_s - [y_p'' + p(t)y_p' + q(t)y_p] =$$

$$= g(t) - g(t) = 0$$

Now, we show that $y_s(t) = y_h(t) + y_p(t)$ is a solution:

$$y_h'' + y_p'' + p(t)(y_h' + y_p') + q(t)(y_h + y_p) =$$

$$= y_h'' + p(t)y_h' + q(t)y_h + [y_p'' + p(t)y_p' + q(t)y_p] =$$

$$= 0 + g(t) = g(t)$$

□

So, to find all solutions of $y''(t) + p(t)y'(t) + q(t)y(t) = g(t)$, we need to find a particular solution y_p and all the solutions of $y''(t) + p(t)y'(t) + q(t)y(t) = 0$, which is called the *associated homogeneous equation*. So, we focus on finding solutions of this equation, leaving, for the moment, the question of particular solutions to a later section. However, we will consider only some special cases for the function g.

By Theorem 6.3.3 and Lemma 6.4.1, we have immediately the following corollary that, though immediate, it is very important for the exercises.

Corollary 6.4.2 *Every solution of the differential equation:* $y''(t) + p(t)y'(t) + q(t)y(t) = g(t)$, *with* p, q *continuous functions in an interval* I, *is given by:*

$$y_p(t) + c_1 y_1(t) + c_2 y_2(t)$$

for suitable constants c_1 *and* c_2, *where:*

- $y_p(t)$ *is a particular solution of the given equation;*

- y_1 *and* y_2 *are solutions of the associated homogeneous equation, with non-zero Wronskian in* I.

Let us see an example.

Example 6.4.3 We want to determine the general solution of the homogeneous equation $y''(t) - 5y'(t) = t$, knowing that $y_1(t) = 5$ and $y_2(t) = e^{5t}$ are solutions of the associated homogeneous equation (see Example 6.3.4) and $y_p(t) = -\frac{t^2}{10} - \frac{t}{25}$ is a particular solution. We leave to the reader the easy check of the last statement. We can, then, immediately write the general solution:

$$y(t) = 5c_1 + c_2 e^{5t} - \frac{t^2}{10} - \frac{t}{25}$$

In the next sections, we will see effective methods for determining solutions of non-homogeneous equations.

The general solution of a second-order linear equation is also necessary for the solution of Cauchy problems.

Example 6.4.4 We want to solve the following initial value problem:

$$\begin{cases} y''(t) - 5y'(t) = t \\ y(0) = 0, \ y'(0) = 1 \end{cases}$$

We proceed first to solve the linear equation, writing the general solution (see the previous example):

$$y(t) = 5c_1 + c_2 e^{5t} - \frac{t^2}{10} - \frac{t}{25}$$

Then, we impose the two initial conditions:

$$y(0) = 5c_1 + c_2 = 0, \qquad y'(0) = 5c_2 - 1/25 = 1$$

Hence, the solution of Cauchy's problem is:

$$y(t) = (1/250)(-25t^2 - 10t + 52e^{5t} - 52)$$

6.5 LINEAR EQUATIONS WITH CONSTANT COEFFICIENTS

We aim to solve linear differential equations with constant coefficients, in other words, equations of the form:

$$ay''(t) + by'(t) + cy(t) = g(t) \qquad (6.9)$$

where the coefficients a, b and c are real numbers, $a \neq 0$. We notice immediately that we can apply to such equations the theory developed so far, taking $p(t) = b/a$ and $q(t) = c/a$. So, by Theorem 6.3.3 and Corollary 6.4.2, we have that, in order to determine the general solution, we need:

- two solutions y_1, y_2 of the associated homogeneous equation, with non-zero Wronskian;

- a particular solution y_p.

The general solution is, then, given by: $y(t) = c_1 y_1(t) + c_2 y_2(t) + y_p(t)$, with c_1, c_2 real constants.

We begin by examining the case in which $g(t) = 0$, i.e., the homogeneous case:

$$ay''(t) + by'(t) + cy(t) = 0 \qquad (6.10)$$

To determine a particular solution, we make the following assumption: we assume that the solution has form $y(t) = e^{\lambda t}$. In order to determine the constant λ, we impose that y is a solution of the equation (6.10).

Substituting into the equation, we get:

$$a\lambda^2 e^{\lambda t} + b\lambda e^{\lambda t} + ce^{\lambda t} = (a\lambda^2 + b\lambda + c)e^{\lambda t} = 0,$$

Hence, since $e^{\lambda t} \neq 0$, we obtain the equation:

$$a\lambda^2 + b\lambda + c = 0 \qquad (6.11)$$

called *auxiliary* or *characteristic* equation, associated with the given homogeneous differential equation. We divide our treatment into three cases, depending on the solutions of (6.11):

1. $b^2 - 4ac > 0$: two distinct real roots, $\lambda_1 \neq \lambda_2$;

2. $b^2 - 4ac = 0$: one real root λ_0;

3. $b^2 - 4ac < 0$: two complex conjugate roots $\alpha \pm i\beta$;

As for the last case, $b^2 - 4ac < 0$: the formula for solving the characteristic equation gives us the solutions

$$\frac{-b \pm \sqrt{b^2 - 4ac}}{2a} = -\frac{b}{2a} \pm \frac{\sqrt{-1}}{2a}\sqrt{|b^2 - 4ac|} = \alpha \pm i\beta$$

Thus, we obtain the two real numbers $\alpha = -b/2a$, $\beta = \sqrt{|b^2 - 4ac|}/2a$, while $i = \sqrt{-1}$ is the *imaginary unit* in complex numbers, whose complete treatment is beyond the scope of the present textbook.

The following theorem is the most important result in the theory of differential equations with constant coefficients, and it is of fundamental importance for solving exercises.

Theorem 6.5.1 *The differential equation:*

$$ay''(t) + by'(t) + cy(t) = 0, \qquad a, b, c \text{ real constants, } a \neq 0$$

has general solution $c_1 y_1(t) + c_2 y_2(t)$ for all real t, where:

1. *if $b^2 - 4ac > 0$, $y_1(t) = e^{\lambda_1 t}$, $y_2(t) = e^{\lambda_2 t}$ with λ_1, λ_2 distinct solutions of the auxiliary equation;*

2. *if $b^2 - 4ac = 0$, $y_1(t) = e^{\lambda_0 t}$, $y_2(t) = te^{\lambda_0 t}$ with λ_0 (unique) solution of the auxiliary equation;*

3. *if $b^2 - 4ac < 0$, $y_1(t) = e^{\alpha t} \cos(\beta t)$, $y_2(t) = e^{\alpha t} \sin(\beta t)$ with $\alpha \pm i\beta$ complex conjugate solutions of the auxiliary equation.*

Proof. Based on our previous discussion, we just have to check, for each case, that y_1 and y_2 are solutions and that their Wronskian is non-zero. Indeed, Theorem 6.3.3 guarantees us that the general solution of the equation $ay''(t) + by'(t) + cy(t) = 0$ is $c_1 y_1(t) + c_2 y_2(t)$.

We show only case (1), leaving the remaining two cases to the reader as an exercise.

We check that the Wronskian of $y_1(t) = e^{\lambda_1 t}$, $y_2(t) = e^{\lambda_2 t}$ is non-zero:

$$W(t) = y_1 y_2' - y_2 y_1' = \lambda_1 e^{\lambda_1 t} e^{\lambda_2 t} - \lambda_2 e^{\lambda_1 t} e^{\lambda_2 t} = (\lambda_1 - \lambda_2) e^{\lambda_1 + \lambda_2}$$

Since $b^2 - 4ac > 0$, the characteristic equation has distinct roots $\lambda_1 \neq \lambda_2$. Hence, $W(t) \neq 0$, for all real t. $\square$

Let us see some examples.

Example 6.5.2 We want to determine the general solution of the homogeneous equation:

$$y''(x) - 3y'(x) + 2y(x) = 0$$

The characteristic equation is $\lambda^2 - 3\lambda + 2 = 0$ and has two solutions $\lambda = 1, 2$. Hence, we have two particular solutions for the differential equation: $y_1 = e^x$, $y_2 = e^{2x}$. By the previous theorem, the general solution is:

$$y = c_1 e^x + c_2 e^{2x}$$

We now look at the solution of a Cauchy problem.

Example 6.5.3 We want to find the solution to the initial value problem:

$$\begin{cases} y''(x) - 2y'(x) - 3y(x) = 0 \\ y(0) = 1 \\ y'(0) = 0 \end{cases}$$

The auxiliary equation is: $\lambda^2 - 2\lambda - 3 = 0$ with solutions $-1, 3$. Hence, the general solution is:

$$y(x) = c_1 e^{-x} + c_2 e^{3x}$$

We now impose the initial conditions:

$$y(0) = c_1 e^{-0} + c_2 e^{3 \cdot 0} = c_1 + c_2 \qquad y'(0) = -c_1 e^{-0} + 3c_2 e^{3 \cdot 0} = -c_1 + 3c_2$$

We have to solve the linear system:

$$\begin{aligned} c_1 + c_2 &= 1 \\ -c_1 + 3c_2 &= 0 \end{aligned}$$

We obtain:

$$c_1 = \tfrac{3}{4} \qquad c_2 = \tfrac{1}{4}.$$

Hence, the solution to the given initial value problem is:

$$y = \tfrac{3}{4} e^{-x} + \tfrac{1}{4} e^{3x}$$

Example 6.5.4 We now want to determine the solution to the initial value problem:

$$\begin{cases} y''(x) - 2y'(x) + 5y(x) = 0 \\ y(0) = 1, \quad y'(0) = 0. \end{cases}$$

The auxiliary equation is given by $\lambda^2 - 2\lambda + 5 = 0$ and has solutions $1 \pm 2i$. Hence, we have the two solutions

$$y_1 = e^x \cos(2x) \qquad y_2 = e^x \sin(2x).$$

and the general solution is:

$$y = c_1 e^x \cos(2x) + c_2 e^x \sin(2x).$$

We impose the initial conditions:

$$1 = y(0) = c_1 \cdot 1 + c_2 \cdot 0 = c_1$$

$$0 = y'(0) = c_1 \cdot (0 + 2) + c_2 \cdot (-2 + 0) = 2c_1 - 2c_2$$

Then, $c_1 = 1$ and $c_2 = -\frac{1}{2}$. Hence, the solution of the given Cauchy problem is:

$$y(x) = \tfrac{1}{2}e^x(2\cos(2x) - \sin(2x))$$

6.6 EQUATIONS WITH CONSTANT COEFFICIENTS: THE GENERAL CASE

We now come to the case of a non-homogeneous equation, that is, of the type:

$$ay''(t) + by'(t) + cy(t) = g(t), \qquad a, b, c \quad \text{constants}, \ a \neq 0 \quad (6.12)$$

with $g(t) \neq 0$. From Theorem 6.3.3, Lemma 6.4.1 and Corollary 6.4.2, we know that, to get the general solution, we need to solve the associated homogeneous equation and take the sum of its general solution and a particular solution $y_p(t)$. Hence, since in the previous section we explained how to find the general solution of homogeneous equations, we just need to find a particular solution of the equation (6.12).

We want to determine $y_p(t)$ using the *method of undetermined coefficients*. This is a very effective method, which, however, allows us to get a solution only for $g(t)$ of the following form: polynomial, exponential or trigonometric (sine and cosine). It is possible to give a variation of this method to obtain the solution for more complicated functions $g(t)$, but we will not examine this more general case here.

The following table tells us how to choose the particular solution $y_p(t)$ for a given $g(t)$:

$g(t)$	$y_p(t)$
$a_0 + a_1 t + \cdots + a_n t^n$	$A_0 + A_1 t + \cdots + A_n t^n$
$ce^{\lambda t}$	$Ce^{\lambda t}$
$a\cos(\omega t) + b\sin(\omega t)$	$A\cos(\omega t) + B\sin(\omega t)$

We observe that the coefficients $a_0, a_1, \ldots, c, \lambda, \omega$ appearing in the expression of $g(t)$ are real numbers, while the coefficients $A_0, A_1, \ldots,$ C, A, B in $y_p(t)$ are variables, which we must determine by imposing that $y_p(t)$ is a particular solution of the given equation: they are the *undetermined coefficients*. It is important to note that, in the expression of $y_p(t)$, the real numbers λ, ω also appear and they are coefficients appearing in the expression of $g(t)$, not undetermined coefficients.

Let us see in practice how to choose a particular solution, using this table. We warn the reader that this method does not necessarily lead us to a particular solution. However, it is very effective in most cases, regarding all the interesting applications in physics and biology.

Example 6.6.1 We want to find the general solution of the equation:

$$y''(x) - 2y'(x) + y(x) = x^2$$

From the table, we see that $y_p(t) = Ax^2 + Bx + C$, from which: $y_p'(x) = 2Ax + B$, $y_p''(x) = 2A$. We substitute in the given equation:

$$(2A) - 2(2Ax + B) + (Ax^2 + Ax + C) = x^2.$$

Hence, $2A - 2B + C = 0$, $-4A + B = 0$, $A = 1$. Then, we find $A = 1, B = 4, C = 6$. We obtain:

$$y_p = x^2 + 4x + 6$$

To find the general solution, we have to solve the associated homogeneous equation:

$$y'' - 2y' + y = 0$$

with auxiliary equation $\lambda^2 - 2\lambda + 1 = 0$, from which $\lambda_0 = 1$. The associated homogeneous equation has general solution:

$$y_h = c_1 e^x + c_2 x e^x$$

The general solution of the given differential equation is the sum of the general solution of the associated homogeneous equation and the particular solution:

$$y = y_h + y_p = c_1 e^x + c_2 x e^x + x^2 + 4x + 6$$

Notice the following important fact: though $g(x)$ contains just one of the powers of x, that is, $g(x) = x^2$, in our application of the method of undetermined coefficients, we have written the polynomial $y_p(x) = Ax^2 + Bx + C$. So, we need to consider all powers of x and not just x^2. Indeed, we have obtained $y_p(x) = x^2 + 4x + 6$, which contains also the term in x and the constant one. Hence, it is wrong to take just $y_p(x) = Ax^2$, because it does not lead us to a particular solution.

Let us now see another example, where the method of undetermined coefficients does not work immediately, but we need a consider a small variation of it.

Example 6.6.2 Let us consider the equation:

$$y''(x) - 2y'(x) - 3y(x) = e^{-x}$$

The associated homogeneous equation has general solution (see Example 6.5.3):

$$y_h(x) = c_1 e^{-x} + c_2 e^{3x}$$

We note that $g(x) = e^{-x}$, hence such $g(x)$ is a solution of the associated homogeneous equation. If we set, according to the table, $y_p(x) = Ae^{-x}$, we immediately see that we get the equation:

$$0 = e^{-x}$$

and we cannot determine the coefficient A.

In these cases, it is useful to multiply by a power of x the particular solution proposed by the table. Hence, we impose $y_p(x) = Axe^{-x}$, so that $y_p'(x) = Ae^{-x} - Axe^{-x}$ and $y_p''(x) = -Ae^{-x} - (Ae^{-x} - Axe^{-x})$. If we substitute in the equation, we get:

$$-2Ae^{-x} + Axe^{-x} - 2(Ae^{-x} - Axe^{-x}) - 3Axe^{-x} = e^{-x}$$

We obtain $A = -1/4$. The general solution of the given equation is:

$$y(x) = c_1 e^{-x} + c_2 e^{3x} - \tfrac{1}{4}xe^{-x}$$

Let us now look at the case where $g(t)$ is given by a sum.

Example 6.6.3 Let us consider the differential equation:

$$y''(t) - 4y(t) = \sin(3t) + e^t$$

We first solve the associated homogeneous equation. The characteristic equation has solutions $\lambda = \pm 2$; hence:

$$y_h(t) = c_1 e^{-2t} + c_2 e^{2t}$$

We now need to determine a particular solution. We impose, according to the table:

$$y_p(t) = A\cos(3t) + B\sin(3t) + Ce^t$$

Since none of the functions appearing in $y_p(t)$ is a solution of the associated homogeneous equation, we expect this choice to work. Substituting in the equation and solving the system, we obtain:

$$A = 0, \quad B = -1/13, \quad C = -1/3$$

from which:

$$y(t) = c_1 e^{2t} + c_2 e^{-2t} - \frac{e^t}{3} - \frac{1}{13}\sin(3t)$$

6.7 SIMPLE HARMONIC MOTION

In this section, we want to examine some applications of the theory developed so far.

Simple harmonic motion is the motion of an *harmonic oscillator*, which is a mechanical system that reacts to a perturbation to its equilibrium state with an acceleration proportional to the displacement, with a constant of proportionality k:

$$F_H = ma = m\frac{d^2x(t)}{dt^2} = -kx(t), \qquad k > 0 \tag{6.13}$$

$F_H = -kx$ is called *Hooke's force* and $k > 0$ is the *constant of the harmonic oscillator*, while $x(t)$ denotes the position.

We are using Newton's law, which tells us that the force F_H is given by the mass m multiplied by the acceleration a. We also recall that the acceleration is the second derivative of the distance as function of time, (see Chapter 3).

Simple harmonic motion can be observed in nature, as the motion of a vibrating spring without friction or other forces acting on it. Let us see an example.

Example 6.7.1 Suppose we have a spring fixed to a wall with a mass of 10 kg attached to it, laying on the floor of a room. Assume the constant $k = 40$ N/m, where N $=$ kg $\cdot$ m/s^2 indicates the unit of measure called *Newton*.

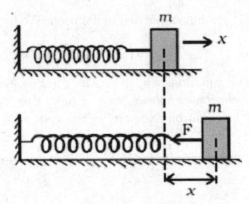

Suppose we displace the mass with respect to the initial equilibrium position by 10 m, that is, $x(0) = 10$ m, with velocity $x'(0) = 5$ m/s. We want to determine the motion of the mass m, assuming no friction with the floor and no other external forces.

We need to solve the following initial value problem:

$$\begin{cases} 10x''(t) = -40x(t) \\ x(0) = 10 \\ x'(0) = 5 \end{cases}$$

We start with the solution of the characteristic equation: $\lambda^2 + 4 = 0$. We get $\lambda = \pm 2i$, and hence, we can write the general solution for the associated homogeneous equation:

$$x(t) = c_1 \cos(2t) + c_2 \sin(2t)$$

By imposing the initial conditions, we finally get the solution of the Cauchy problem:

$$x(t) = \frac{5}{2}[\sin(2t) + 4\cos(2t)]$$

The example shows us that the equation (6.13) corresponds to a *periodic* motion, that is, a motion repeating itself after a certain interval of time called *period*. In the above example, the argument of sine and cosine is $2t$, and so, the period is $T = 2\pi/2 = \pi$, i.e., $x(t) = x(t + \pi)$ for each real t.

Let us now see from a more theoretical point of view, the calculations we did in the previous example. We can solve the equation (6.13), according to the theory developed so far:

$$mx''(t) = -kx(t)$$

We immediately obtain the characteristic equation: $m\lambda^2 + k = 0$, with solutions $\lambda = \pm i\sqrt{k/m}$ leading to the general solution:

$$x(t) = c_1 \cos(\sqrt{k/m}\,t) + c_2 \sin(\sqrt{k/m}\,t)$$

We define $\omega := \sqrt{k/m}$ the *natural frequency* of the harmonic oscillator, while, from elementary trigonometry, we have that the period is given by $T = 2\pi/\omega$.

6.8   HARMONIC MOTION WITH EXTERNAL FORCE

We now want to examine the more complicated situation, when the harmonic oscillator is also subject to an external force.

Let us start with a simple example, of interest in physics. Suppose we have a spring attached to the ceiling of a room, as in the figure below. At the contrary, we suspend a mass, and we give it an initial displacement and velocity, which determine its motion. In this case, the equation (6.13) changes, as we need to take into account gravity $F_G = mg$ ($g = 9.8$ m/s^2 acceleration of gravity):

$$m\frac{d^2x(t)}{dt^2} = -kx(t) + mg \tag{6.14}$$

The motion of the mass is expressed by the solution $x(t)$ of the differential equation (6.14), given the initial conditions $x(0)$ and $x'(0)$, which are, respectively, the displacement and the initial velocity imparted to the mass m.

Given this equation, we can determine the spring constant k, measuring the displacement x_0 at equilibrium, that is, how much the spring extends once the mass m is attached to it, before imparting to the mass the initial displacement and velocity. To determine k, knowing x_0, we hence apply the equation (6.14), with total force equal to zero (we are at equilibrium):

$$mg - kx_0 = 0 \quad \Longrightarrow \quad k = mg/x_0 \qquad (6.15)$$

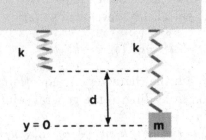

Once we have determined k as above, we can compute the equation in a simpler form by setting $y = x - x_0$. Recalling that x_0 is constant, hence its derivative is zero, and keeping in mind (6.15), we have:

$$m(x - x_0)'' = mx'' = -kx + mg = -kx + kx_0 \quad \Longrightarrow \quad m(x - x_0)''$$
$$= -k(x - x_0) \qquad (6.16)$$

that is:

$$my''(t) = -ky(t) \qquad (6.17)$$

Equation (6.17) gives us the dynamics of an harmonic oscillator, that is, the position $y(t)$ of the mass m with respect to the equilibrium position, with no friction, taking into account only the force of gravity. Formally, we see that this is the same equation we have for simple harmonic motion (i.e. with no external forces). This is due to the fact that the force of gravity *does not* depend on the time. Because of this, through a simple change of variables (6.16), we can transform the equation (6.14) in the simple harmonic oscillator equation (6.17).

Let us see a concrete example of application.

Example 6.8.1 Suppose we have a mass of 2 kg that, when attached to a vertical spring, produces a displacement of 49 cm. Then, because of (6.15), we immediately have:

$$k = mg/x_0 = (2\text{kg} \times 9.8\text{m/s}^2)/0.49\text{m} \cong 40\text{kg/s}^2 = 40 \text{ N/m}$$

(N=1 kg m/s^2 is the Newton unit).

We write the differential equation (6.13):

$$2y''(t) = -40y$$

which gives the displacement $y(t)$ of the mass. Using the methods studied so far, we know that, to solve such an equation, we need to find the roots of the characteristic equation:

$$2\lambda^2 + 40 = 0 \quad \Longrightarrow \quad \lambda = \pm 2i\sqrt{5}.$$

The general solution is:

$$y(t) = c_1 \cos(2\sqrt{5}t) + c_2 \sin(2\sqrt{5}t)$$

To determine the constants c_1 and c_2, we need know the initial conditions. For example, assume the spring is at the equilibrium position and not subject to other forces, except the force of gravity and the Hooke force, which compensate each other at equilibrium. So, if the initial conditions are $y(0) = y'(0) = 0$, we have $y(t) = 0$ for every t, i.e., the mass m remains at the equilibrium.

If we displace the mass 1 m downward and we impart a downward velocity of 0.5 m/s, we get the initial conditions:

$$y(0) = 1, \qquad y'(0) = 0.5$$

Substituting in the expression of $y(t)$ and $y'(t)$, we have:

$$y(t) = \cos(2\sqrt{5}t) + \frac{1}{4\sqrt{5}} \sin(2\sqrt{5}t)$$

which gives the position of the mass at time t, with respect to the equilibrium position.

More generally, in the study of the harmonic oscillator, we can also have an external force, which is not necessarily the force of gravity, and can depend on time. In this case, Hooke's law becomes:

$$mx''(t) = -kx(t) + F(t) \tag{6.18}$$

where $F(t)$ represents the external force.

Let us see an example.

Example 6.8.2 Suppose we have a spring with constant $k = 2$ N/m, with one end fixed to a vertical wall. At the other end, we have a mass of 1 kg, lying on the floor and subject to a periodic force $F(t) = \sin(3t)$. We want to determine the motion of the mass m, assuming that it is initially at rest, that is, $x(0) = 0$, $x'(0) = 0$.

We set up Hooke's law (6.18), where we take into account a external force:

$$mx''(t) = -kx(t) + \sin(3t)$$

So, the differential equation to be solved is:

$$x''(t) + 2x(t) = \sin(3t)$$

with the initial conditions $x(0) = 0$, $x'(0) = 0$.

We first determine the solutions of the associated homogeneous equation. The characteristic equation $\lambda^2 + 2 = 0$ has a solution $\pm i\sqrt{2}$, hence

$$x(t) = c_1 \cos(\sqrt{2}t) + c_2 \sin(\sqrt{2}t)$$

We now look for a particular solution, using the method of undetermined coefficients: $x_p(t) = A\sin(3t) + B\cos(3t)$. Substituting $x_p(t)$ in $x''(t) + 2x(t) = \sin(3t)$, we obtain $x_p(t) = -(1/7)\sin(3t)$. Hence, the general solution of the given equation is:

$$x(t) = c_1 \cos(\sqrt{2}t) + c_2 \sin(\sqrt{2}t) - (1/7)\sin(3t)$$

We impose the initial conditions $x(0) = 0$, $x'(0) = 0$ and we obtain:

$$x(t) = (1/14)(3\sqrt{2}\sin(\sqrt{2}t) - 2\sin(3t))$$

We observe the following obvious fact: if we have no external force and we impose the initial conditions $x(0) = 0$, $x'(0) = 0$, that is no displacement and initial velocity, then the motion of the oscillator harmonic is $x(t) = 0$; the mass will remain in its initial the position. However, if we have an external force, as in this case, then, even if we do not have an initial displacement and an initial velocity, the harmonic oscillator has a non-zero time evolution $x(t)$.

6.9 DAMPED HARMONIC MOTION

In real harmonic oscillators, the friction or damping of the medium slows down the motion of the system. Hence, such a system is called *damped harmonic oscillator*. It follows Hooke's law (6.18), where $F(t)$

is the friction force. This force is always opposing the motion, and it is proportional to the velocity $y'(t)$, according to the damping coefficient $\mu > 0$. We hence have:

$$my''(t) = -ky(t) - \mu y'(t)$$

So, the differential equation that determines the damped harmonic motion is given by:

$$my'' + \mu y' + ky = 0$$

We distinguish three cases, as we did in Section 6.3, depending on the sign of the discriminant $\Delta = \mu^2 - 4km$ of the characteristic equation $m\lambda^2 + \mu\lambda + k = 0$.

- **Overdamping:** $\Delta > 0$: We have distinct solutions λ_1, λ_2.

  The general solution of the differential equation is:

  $$y(t) = c_1 e^{\lambda_1 t} + c_2 e^{\lambda_2 t}$$

- **Critical damping:** $\Delta = 0$: We have a unique solution $\lambda = \lambda_0$.

  The general solution of the differential equation is:

  $$y(t) = c_1 e^{\lambda_0 t} + c_2 t e^{\lambda_0 t}$$

- **Underdamping:** $\Delta < 0$: We have solutions $\lambda = \alpha \pm i\beta$.

  The general solution of the differential equation is:

  $$y(t) = e^{\alpha t}[c_1 \cos(\beta t) + c_2 \sin(\beta t)]$$

Let us see an example.

Example 6.9.1 A mass of 2 kg is attached to a spring suspended from the ceiling, with constant $k = 150$ kg/s^2 and with damping constant $\mu = 10$ kg/s. Suppose the mass is subject to a downward velocity of 5 cm/s. We want to determine the position of the mass, with respect to the equilibrium position, as a function of the time. We can immediately set up the differential equation:

$$2y''(t) + 10y'(t) + 150y(t) = 0$$

The solutions of the characteristic equation $2\lambda^2 + 10\lambda + 150 = 0$ are $(1/2)(-5 \pm 5i\sqrt{11})$. We are in the case of *underdamping*. The general solution is:

$$y(t) = e^{-5t/2}[c_1 \sin(5\sqrt{11}t/2) + c_2 \cos(5\sqrt{11}t/2)]$$

Imposing the initial conditions, corresponding to no initial displacement and an initial velocity of 5 cm/s (conventionally the positive direction is downward), we get:

$$y(t) = \frac{2}{\sqrt{11}}e^{-5t/2} \sin(5\sqrt{11}t/2)$$

which gives the position of the mass at all times.

Let us see another example, important for applications in physics.

Example 6.9.2 Suppose we have a mass of 10 kg attached to a vertical spring with constant $k = 90$ kg/s^2. We want to determine the damping constant μ, so that the motion is critically damped. This is particularly useful in concrete applications. For example, when designing shock absorbers in vehicles: we want just one oscillation to cushion the shocks, without further movements.

The differential equation to solve is:

$$10y''(t) + \mu y'(t) + 90y(t) = 0$$

The associated characteristic equation is given by: $10\lambda^2 + \mu\lambda + 90 = 0$. To have critical damping, we must have a unique solution; hence, we impose that $\mu^2 - 3600 = 0$ and we get $\mu = 60$ kg/s.

The general solution of the equation is:

$$y(t) = c_1 e^{-3t} + c_2 e^{-3t}t$$

Suppose now we impose a displacement of $y(0) = 1$ cm with velocity $y'(0) = 5$ cm/s. We get the solution:

$$y(t) = e^{-3t}(8t + 1)$$

whose graph is given by:

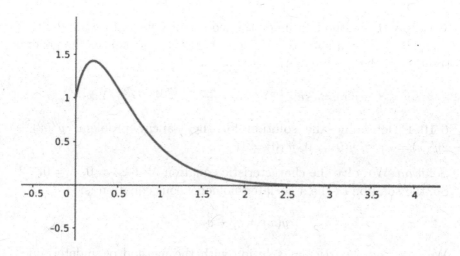

6.10 EXERCISES WITH SOLUTIONS

6.10.1 Solve the differential equation: $y''(x) + 3y'(x) + 4y(x) = 0$.

Solution. We first solve the characteristic equation: $\lambda^2 + 3\lambda + 4 = 0$, $\lambda = (1/2)[-3 \pm i\sqrt{7}]$. The general solution is:

$$y(x) = e^{-3x/2}[c_1 \sin(\sqrt{7}x/2) + c_2 \cos(\sqrt{7}x)/2)]$$

6.10.2 Solve the differential equation with initial conditions: $2y''(x) - 2y'(x) + 5y(x) = 0$, $y(0) = 3$, $y'(0) = -2$.

Solution. First of all, we solve the differential equation. The characteristic equation $2\lambda^2 - 2\lambda + 5 = 0$ has solutions $\lambda = (1/2)[1 \pm i\sqrt{3}]$. Then:

$$y(x) = e^{x/2}[c_1 \sin(3x/2) + c_2 \cos(3x/2)]$$

Imposing the initial conditions, we obtain:

$$y(x) = \frac{1}{3}e^{x/2}[9\cos(3x/2) - 7\sin(3x/2)]$$

6.10.3 Solve the differential equation: $y''(x) + 2y'(x) - 4y(x) = e^{2x}$.

Solution. Let us compute the roots of the characteristic equation $\lambda^2 + 2\lambda - 4 = 0$, $\lambda = -1 \pm \sqrt{5}$. We obtain the general solution of the associated homogeneous equation:

$$y(x) = c_1 e^{(-1-\sqrt{5})x} + c_2 e^{(-1+\sqrt{5})x}$$

We apply the method of undetermined coefficients and we impose that $y_p(x) = Ae^{2x}$ is a solution. We get $A = 1/4$, and hence, the general solution of the given equation is:

$$y(x) = c_1 e^{(-1-\sqrt{5})x} + c_2 e^{(-1+\sqrt{5})x} + e^{2x}/4$$

6.10.4 Determine the solution for the Cauchy problem: $y''(x) + 2y'(x) = xe^{2x}$, $y(0) = 0$, $y'(0) = 1$.

Solution. We solve the characteristic equation $\lambda^2 + 2\lambda = 0$, $\lambda = 0, -2$. So, the solution of the associated homogeneous equation is:

$$y(x) = c_1 + c_2 e^{-2x}$$

We now find a particular solution with the method of undetermined coefficients. We substitute $y_q(x) = (A + Bx)e^{2x}$ in the given equation and we find:

$$y(x) = \frac{1}{32} e^{2x}(4x - 3)$$

Hence, the general solution is:

$$y(x) = c_1 + c_2 e^{-2x} + \frac{1}{32} e^{2x}(4x - 3)$$

Imposing the initial conditions, we get:

$$y(x) = \frac{1}{32}[e^{2x}(4x - 3) - 17e^{-2x} + 20]$$

6.10.5 A 2 kg weight extends a spring by 5 m. We then impart an initial upward velocity of 10 m/s. Determine the position of the mass $y(t)$, with respect to the equilibrium, as a function of time.

Solution. We can immediately find the constant $k \cong (2 \times 9.8)/5 \cong 4$. The harmonic oscillator equation is $y''(t) + 4y(t) = 0$ and has general solution:

$$y(t) = c_1 \cos(2t) + c_2 \sin(2t)$$

We impose the initial conditions $y(0) = 0$, $y'(0) = -10$. We put the negative sign, because the displacement is upward. Conventionally, we take the displacement as positive if downward (see figure in Section 6.8). We obtain:

$$y(t) = -5 \sin(2t)$$

6.10.6 Suppose we have a mass of 2 kg hanging from a spring attached to the ceiling of a room with. $k = 50$ kg/s^2. If the damping constant is 20 kg/s and $y(0) = 1$, $y'(0) = 5$, determine the position of the mass as time varies. Moreover, determine if the motion is overdamped, underdamped or critically damped.

Solution. The differential equation we need to solve is:

$$2y''(t) + 20y'(t) + 50y(t) = 0$$

We write the characteristic equation: $2\lambda^2 + 20\lambda + 50 = 0$, which has a unique solution $\lambda = -5$. We have a damped harmonic oscillator with critical damping. The general solution is given by:

$$y(t) = c_1 e^{-5t} + c_2 e^{-5t} t$$

If we impose the initial conditions, we get:

$$y(t) = e^{-5t}(10t + 1)$$

6.11   SUGGESTED EXERCISES

6.11.1 For any of the following differential equations find the general solution.

1. $y'' - 2y' - 8y = 0$

2. $2y'' - y' - y = 0$

3. $y''(t) - 2y'(t) + y(t) = 0$

4. $y''(t) + 2y(t) = 0$

6.11.2 Solve the following Cauchy problems:

- $y'' - 2y' - 8y = 0$, $y(0) = 0$, $y'(0) = 1$.

- $2y'' - y' - y = 0$, $y(0) = -1$, $y'(0) = 1$.

- $y''(t) + 2y(t) = 0$, $y(0) = 0$, $y'(0) = 1$.

6.11.3 For each of the differential equations find a particular solution and then the general solution.

- $y'' - 2y' - 8y = e^{2t}$

- $2y'' - y' - y = 4\sin(t)$

- $y''(t) - 2y'(t) + y(t) = t^2$

- $y''(t) + 2y(t) = t^2$

6.11.4 Solve the following problems with initial conditions:

- $y'' - 2y' - 8y = e^{2t}$, $y(0) = 0$, $y'(0) = 1$.

- $2y'' - y' - y = 5\sin(t)$, $y(0) = -1$, $y'(0) = 1$.

- $y''(t) + 2y(t) = t^2$, $y(0) = 0$, $y'(0) = 1$.

6.11.5 Solve the following problems with initial conditions:

1. $y'' - 3y' + 2y = 0$, $y(0) = 1$, $y'(0) = 0$.

2. $y'' - 2y' + 2y = 0$, $y(0) = 1$, $y'(0) = 0$.

3. $y'' - y' + 2y = 0$, $y(0) = 1$, $y'(0) = 0$.

4. $y'' - 3y' + 2y = 3e^x$, $y(0) = 1$, $y'(0) = 0$.

5. $y'' - 3y' + 2y = 4\cos(x)$, $y(0) = 0$, $y'(0) = 1$.

6. $y'' - y' - 2y = 4x^2 - e^x$, $y(0) = 1$, $y'(0) = 0$.

6.11.6 Solve the following problems with initial conditions:

1. $y'' - 4y = e^{2x}$, $y(0) = 0$, $y'(0) = 1$.

2. $y'' + y = 4\sin(x)$, $y(0) = -1$, $y'(0) = 1$.

3. $y''(x) - y(x) = x^2 + e^x$, $y(0) = 0$, $y'(0) = 1$.

6.11.7 A weight of 200 g is attached to a spring and extends the spring by 5 cm. Determine the equation of motion $x(t)$ of the spring if the weight is furtherly extended, moving it from its position of equilibrium, by 10 cm.

6.11.8 A weight of mass 1 kg is attached to a vertical spring with constant $k = 1$ N/m and friction coefficient $\mu = 1$ kg/s. If we impart to the spring a downward velocity of 2 m/s, determine the position $y(t)$ of the mass as a function of time. Determine a value of the time t for which the velocity becomes zero.

6.11.9 A mass of 1 kg extends a spring 20 cm. If we impart to the spring an upward velocity of 3 m/s, assuming that there is a damping force equal to 14 times the speed, determine the position $x(t)$ of the mass as a function of time.

6.11.10 A weight of 2 kg is suspended from a spring, producing a displacement of 28 cm.

    a. What is the constant of the spring? What is the period of oscillation?

    b. The spring is then moved further downward by 4 cm. Find the function that describes the position of the mass as function of time.

    c. Determine the position of the spring at time $t = 2$ in seconds.

6.11.11 Check that $te^{\lambda_0 t}$ is a solution of the differential equation $ay'' + by' + cy = 0$, in the notation of Theorem 6.5.1. Also check that the Wronskian of $e^{\lambda_0 t}$ and $te^{\lambda_0 t}$ is not equal to zero for all real t.

6.11.12 Check that $e^{\alpha t}\sin(\beta t)$ and $e^{\alpha t}\cos(\beta t)$ are both solutions of the differential equation $ay'' + by' + cy = 0$, in the notation of Theorem 6.5.1. Also check that their Wronskian is different from zero for all $t \in \mathbb{R}$.

    [Note: Euler's formula states that $e^{it} = \cos(t) + i\sin(t)$.]

6.12 APPENDIX: LINEAR SYSTEMS

In this section, we want to give a condition for the existence of solutions for a generic linear system of the type.

$$\begin{cases} a_{11}x + a_{12}y = b_1 \\ a_{21}x + a_{22}y = b_2 \end{cases} \tag{6.19}$$

We can assume that $a_{11} \neq 0$ up to rearranging the equations (if both a_{11} and a_{21} were zero, we would not have the unknown x). So we write:

$$\begin{cases} x = (1/a_{11})(b_1 - a_{12}y) \\ a_{21}(1/a_{11})(b_1 - a_{12}y) + a_{22}y = b_2 \end{cases}$$

From which we obtain that, if $a_{11}a_{22} - a_{12}a_{21} \neq 0$:

$$\begin{cases} x = \frac{b_1 a_{22} - a_{12} b_2}{a_{11} a_{22} - a_{12} a_{21}} \\ \\ y = \frac{b_2 a_{11} - a_{21} b_1}{a_{11} a_{22} - a_{12} a_{21}} \end{cases}$$

Hence, if $a_{11}a_{22} - a_{12}a_{21} \neq 0$, then the system admits solution and this solution is unique, because we can determine it explicitly and uniquely. Conversely, if $a_{11}a_{22} - a_{12}a_{21} = 0$, then we obtain that $a_{11}x + a_{12}y$ is a multiple of $a_{21}x + a_{22}y$. The system admits solution if and only if b_1 is multiple, according to the same constant, of b_2, otherwise the system has no solution.

Given a matrix of a system:

$$A = \begin{pmatrix} a_{11} & a_{12} \\ a_{21} & a_{22} \end{pmatrix}$$

we define the *determinant* of A as $\det(A) = a_{11}a_{22} - a_{12}a_{21}$.

Consequently, given a linear system (6.19) with two equations and two unknowns, we have only three possibilities expressed by the following proposition, which we proved in our previous discussion.

Proposition 6.12.1 *Given the linear system:*

$$\begin{cases} a_{11}x + a_{12}y = b_1 \\ a_{21}x + a_{22}y = b_2 \end{cases}$$

Then:

1. *If* $\det(A) = a_{11}a_{22} - a_{12}a_{21} \neq 0$, *then the system admits a unique solution.*

2. *If* $\det(A) = a_{11}a_{22} - a_{12}a_{21} = 0$, *then:*

   - *the system does not admit solutions or*
   - *the system admits infinitely many solutions.*

Note that if $b_1 = b_2 = 0$, then we have only two possibilities: the first and the third. Such systems are called *homogeneous*.

Looking at the conditions that we have obtained above, we have already proved the following proposition.

Proposition 6.12.2 *Consider the homogeneous linear system:*

$$\begin{cases} c_{11}x + c_{12}y = 0 \\ c_{21}x + c_{22}y = 0 \end{cases} \tag{6.20}$$

This system admits a solution different from $x = y = 0$ if and only if the determinant of matrix:

$$C = \begin{pmatrix} c_{11} & c_{12} \\ c_{21} & c_{22} \end{pmatrix}$$

$det(C) = c_{11}c_{22} - c_{12}c_{21}$ *is equal to zero.* ·

CHAPTER 7

Elementary Statistics

7.1   POPULATIONS AND VARIABLES

In statistics, we call *statistical population* (or simply *population*) a set that we want to study, via some experiments or measures. For example, the students of an high school is a population, and we may want to study the performance, in term of grades, in one or more subjects. So, in statistics, once we choose a population, we must also assign characteristics we are interested in. For example, in the population of students mentioned above, characteristics can also be, in addition to grades, height, weight, eye color and so on, depending on the questions we are interested in. Of course, many of the characteristics are not expressed by numerical values and can be unrelated with each other. The main purpose of elementary statistics is to understand how the characteristics of the same population can be correlated. Let us formalize some of the intuitive concepts introduced so far.

Definition 7.1.1 We call *variable* a characteristic assigned to each element of a given statistical population. We define *statistical variable* a set of values for a given variable on the population or on a *sample* of it, that is, a subset of the population where we can give a reasonable estimate of the characteristic that we want to measure. A statistical variable is then identified by a N-tuple of values $Y = (y_1, y_2, \ldots, y_N)$, where N is the number of items in the sample.

In the example of the population consisting of students of a high school, we can consider as variables, for example, the height, weight, grades and eye color of students. The values of these variables,

 DOI: 10.1201/9781003343288-7

measured for each student, are statistical variables. An effective sample for the variable "eye color" is the subset consisting of male students. The same subset is not a good sample, if we want to take into consideration the variable "height", since the height of men is, on average, greater than that of women.

Let us see another example.

Example 7.1.2 The test scores of a sample consisting of 15 students, all enrolled in the same class, are described by the statistical variable:

$$Y = (60, 80, 92, 100, 83, 84, 96, 74, 63, 80, 100, 90, 75, 74, 92)$$

where, in this case, $N = 15$ is the number of students in the sample.

The same variable measured on different samples of the same population may give different statistical variables. For example, if the variable is a score, as in the previous example, the collected data giving the statistical variable can differ a lot when taking different samples, that is, different groups of students.

7.2 ABSOLUTE FREQUENCIES AND PERCENTAGES

When we have a large sample to describe a given variable on a population, it is convenient to write only the *distinct* values and report, for each value, how many times it appears in the statistical variable. Such information is essential, when we want to study phenomena in which we have a large sample with repeated values, for example, the eye color in the high school student population.

Definition 7.2.1 Let $Y = (y_1, y_2, \ldots, y_N)$ be a statistical variable. We define *absolute frequency* of a given value as the number of times it appears in the expression of the statistical variable.

Let us see an example, where we express absolute frequencies in a table, called the *frequency table*.

Example 7.2.2 Let us go back to the Example 7.1.2. Consider the statistical variable, reporting the final grades for a sample of 15 students:

$$Y = (60, 80, 92, 100, 83, 84, 96, 74, 63, 80, 100, 90, 75, 74, 92)$$

We now express the same statistical variable, using only distinct values and absolute frequencies, through the following table, called the frequency table:

Value	60	80	92	100	83	84	96	74	63	90	75
Absolute Freq.	1	2	2	2	1	1	1	2	1	1	1

Clearly, the sum of all absolute frequencies gives the sample size N:

$$N = 1 + 2 + 2 + 2 + 1 + 1 + 1 + 2 + 1 + 1 + 1 = 15.$$

Definition 7.2.3 Let Y be a statistical variable and f be the absolute frequency of the value z. We define *relative frequency* of the value z as the quotient f/N, where N is the number of elements in the sample. The *percentage frequency* is obtained by multiplying the relative frequency by 100.

Naturally, the sum of all relative frequencies is equal to 1, while the sum of all percentage frequencies is equal to 100.

Example 7.2.4 In the Examples 7.1.2 and 7.2.2 regarding the statistical variable "grade" for a sample of 15 students, we obtain the following frequency tables:

Value	60	80	92	100	83	84	96	74	63	90	75
R. F.	$0.0\bar{6}$	$0.1\bar{3}$	$0.1\bar{3}$	$0.1\bar{3}$	$0.0\bar{6}$	$0.0\bar{6}$	$0.0\bar{6}$	$0.1\bar{3}$	$0.0\bar{6}$	$0.0\bar{6}$	$0.0\bar{6}$
P. F.	$6.\bar{6}$	$13.\bar{3}$	$13.\bar{3}$	$13.\bar{3}$	$6.\bar{6}$	$6.\bar{6}$	$6.\bar{6}$	$13.\bar{3}$	$6.\bar{6}$	$6.\bar{6}$	$6.\bar{6}$

where R. F. denotes the relative frequency, while P. F. denotes the percentage frequency.

Sometimes it is useful to group the values a variable can take in *classes*, represented by continuous intervals.

Example 7.2.5 A sample consisting of 300 biotechnology students takes a final exam and the grades, ranging between 18 and 30, are recorded in a statistical variable. Instead of writing all 300 grades, we report the information dividing the interval [18,30] into subintervals (the *classes*) and reporting the number of grades in each subinterval:

Classes	[18, 20]	[20, 22]	[22, 24]	[24, 26]	[26, 28]	[28, 30]
Abs. Freq.	80	55	65	40	35	25
Rel. Freq.	$0.2\bar{6}$	$0.18\bar{3}$	$0.21\bar{6}$	$0.1\bar{3}$	$0.11\bar{6}$	$0.08\bar{3}$

Notice that, with a slight abuse of terminology, we speak of absolute (or relative) frequency of an interval of values and not just one value.

From the table, we can immediately see, for example, that 40 students obtained a grade between 24 and 26 and represent approximately 13% of the given sample.

We can also see that 100 students, that is about 33% of the sample, achieved a grade greater or equal to 24. This information can be quickly obtained by summing the absolute frequencies, for the classes corresponding to the intervals: [24, 26], [26, 28], [28, 30].

7.3   GRAPHICAL REPRESENTATION OF DATA

In this section, we explain how to use graphs to represent data available in tables. We prefer to understand each type of graph through examples, leaving to the student the easy generalization to all different possible cases.

Example 7.3.1 Suppose we measure the height of a sample of students in a middle school and we write the following frequency table:

Height	157	160	165	173	168	176	184
Abs. Freq.	1	2	6	4	3	3	1

We can effectively express this information through a *column diagram* (or column chart), where we put on the x-axis the height in centimeters, while in the y-axis the number of students with a certain height.

Column diagram:

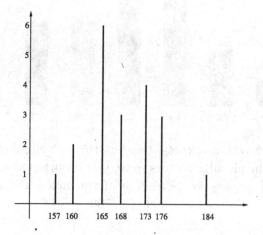

We can express the same data using a *frequency polygon*, obtained by connecting each column tip to the next by a segment.

Frequency polygon:

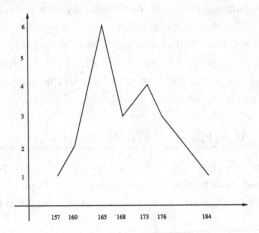

Similarly, we can also use a *bar graph* to represent the same data. It is very similar to the column graph examined above.

Bar graph:

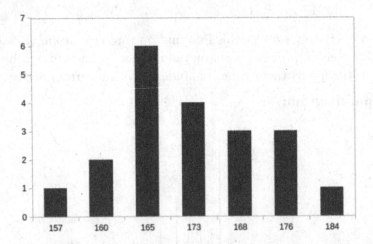

In the *pie charts*, we express the data through circular sectors of the same circle. The circular sectors must be proportional to the absolute frequencies. If q_i are the percentage frequencies, the corresponding circular sectors will have an angle $(q_i/100) \times 360^o$.

Pie chart:

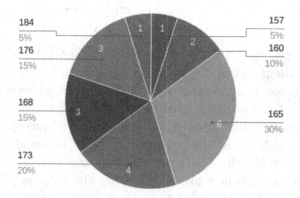

We conclude with the *histograms*, which represent a small variation of the column or bar graphs and are widely used. Hence, an histogram represents the values in a statistical variable, with their frequencies. In the example under consideration, we look at height intervals of 5 cm.

Histogram:

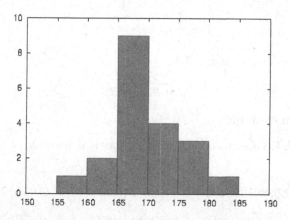

7.4 MODE, AVERAGE AND MEDIAN

When a population is described by a large amount of data, we need to not only organize them via statistical variables but also define some quantities associated with them that can help us to understand better the meaning of our data. The *mode* is a key concept, and it represents the discrete analog of the idea of maximum, which we studied in Chapter 3 for differentiable functions.

Definition 7.4.1 We define the *mode* of a statistical variable as the value occurring with the highest frequency. If the values are grouped into classes, we call *mode class* as the class with the highest frequency.

In the Example 7.2.5, the mode class is $[18, 20]$. If we look at the graphs and diagrams of the previous section, we can determine the mode at a glance: it is the value of the statistical variable corresponding to the peak of the polygon or bar graphs. In the Example 7.3.1, the mode is equal to 165 and occurs with absolute frequency equal to 6. Since we are dealing with discrete variables, that is, variables assuming only a finite set of values, it is not possible to compute their maxima and minima through derivatives. However, the concept of mode and its immediate calculation via graphs help us to solve the important problem of determining the maximum in our discrete setting.

We now come to the very important concepts of arithmetic mean, geometric mean and median.

Definition 7.4.2 Let $Y = (y_1, \ldots, y_N)$ be a statistical variable. We call *arithmetic mean* or more simply *mean or average* of Y:

$$\bar{y} = \frac{1}{N}(y_1 + \cdots + y_N)$$

We define *geometric mean* of Y:

$$\sqrt[N]{y_1 \cdot y_2 \cdot \cdots \cdot y_N}$$

Let us see an example.

Example 7.4.3 Let us consider the statistical variable of the Example 7.1.2

$$Y = (60, 80, 92, 100, 83, 84, 96, 74, 63, 80, 100, 90, 75, 74, 92)$$

regarding the final grades of a sample of 15 students. The arithmetic mean is given by:

$$\bar{y} = \tfrac{1}{15}(60 + 80 + 92 + 100 + 83 + 84 + 96 + 74 + 63 + 80 + 100 + 90$$

$$+ 75 + 74 + 92) = 1243/15 \cong 82.86$$

The geometric mean is given by:

$$\sqrt[15]{1243} \cong 1.608$$

Sometimes, it is useful to compute an average, where each value is *weighed*, that is, multiplied by a positive real number called *weight*.

Definition 7.4.4 Given a statistical variable $Y = (y_1, \ldots, y_N)$, we define *weighted average* of Y:

$$\frac{w_1 y_1 + \cdots + w_N y_N}{w_1 + \cdots + w_N}$$

The positive real numbers $w_1, \ldots, w_N$ are called *weights*.

Let us see a concrete example of this important average.

***Example* 7.4.5** Typically, in the Italian university system, to compute the grade average for a student, each final grade is weighted according to the credits pertaining to that final exam. Let us see in a concrete case what this amounts to. Suppose a student has achieved the grades shown in the table together with the credits for each class:

Exam	Grade	Credits
Biology 1	30	6
Mathematics	21	10
English	30	3

Let us compute both the arithmetic and weighted means.

$$\text{Arithmetic mean} = \frac{30 + 21 + 30}{3} = 27,$$

$$\text{Weighted mean} = \frac{30 \cdot 6 + 21 \cdot 10 + 30 \cdot 3}{6 + 10 + 3} \cong 25.26$$

We immediately see that the exam with a greater number of credits *weighs* more in the calculation of the weighted average, that is, it has a greater impact on the weighted mean.

We conclude this section with the important concept of *median*.

Definition 7.4.6 The *median* of a statistical variable $Y = (y_1, \ldots, y_N)$, where the values are sorted in ascending order, i.e., $y_1 \leq \cdots \leq y_N$, is defined as follows.

- If N is odd, the median is $y_{\frac{N+1}{2}}$. For example, the median of $Y = (2, 5, 8, 20, 33)$ is the value $y_3 = 8$.

- If N is even, the median is equal to the arithmetic mean of the two values $y_{\frac{N}{2}}$, $y_{\frac{N}{2}+1}$. For example, the median of $Y = (2, 5, 8, 20)$ is given by $(y_2 + y_3)/2 = 6.5$.

We note that, in the case of a sample with an even number of elements, the median *is not* one of the values of the statistical variable.

The mean and median look very similar; however, it is important to understand that these two numbers can also be very different, if the statistical variable is associated to a sample with many extreme values. In the Example 7.4.5, we immediately see that the median is 30, while the arithmetic mean is 27. We are going to see, in the next section, more about this, via the concept of variance and standard deviation.

7.5 VARIANCE AND STANDARD DEVIATION

The variance of a statistical variable is a fundamental concept in statistics, as it expresses the "variability" of the data we have available. It measures how much information in a sample, that is, the statistical variable of that sample, is located near the average of the statistical variable values. Roughly, we have that a small variance corresponds to values clustered near their average, while a large variance corresponds to values scattered in a wider range around it.

Let us see the rigorous definition, followed by some concrete examples.

Definition 7.5.1 Let $Y = (y_1, \ldots, y_N)$ be a statistical variable with arithmetic mean $\overline{y}$. We define *variance* of Y as the average of $(y_i - \overline{y})^2$ for i between 1 and N. We define *standard deviation* the square root of the variance.

Some authors write in the denominator $N - 1$ instead of N, but we will use the formula with N as the denominator. We denote, respectively, with σ and σ^2 the standard deviation and the variance of a statistical variable:

$$\sigma^2 = \tfrac{1}{N}[(y_1 - \overline{y})^2 + \cdots + (y_N - \overline{y})^2]$$

$$\sigma = \tfrac{1}{\sqrt{N}} \sqrt{(y_1 - \overline{y})^2 + \cdots + (y_N - \overline{y})^2}$$

We also call $(y_i - \overline{y})^2$ *square deviation*.

The fundamental property of variance is clear from its definition: the smaller the variance, the closer the data to the value of the arithmetic mean. Indeed, if all values in the statistical variable are equal to the arithmetic mean, the variance is zero.

We now look at an important property of the variance.

Proposition 7.5.2 *The variance of a statistical variable is the difference between the arithmetic mean of the squares of the values and the square of their arithmetic mean:*

$$\sigma^2 = \overline{y^2} - \overline{y}^2$$

Proof. This is just a calculation:

$$
\begin{aligned}
\sigma^2 &= \tfrac{1}{N}\left[(y_1 - \overline{y})^2 + \cdots + (y_N - \overline{y})^2\right] = \\
&= \tfrac{1}{N}\left[(y_1^2 - 2\overline{y}y_1 + \overline{y}^2) + \cdots + (y_N^2 - 2\overline{y}y_N + \overline{y}^2)\right] = \\
&= \tfrac{1}{N}(y_1^2 \cdots + y_N^2) - 2\overline{y}\tfrac{1}{N}(y_1 + \ldots y_N) + \overline{y}^2 \\
&= \overline{y^2} - \overline{y}^2
\end{aligned}
\tag{7.1}
$$

$\square$

Now let us see a concrete example.

Example 7.5.3 Suppose we weigh ten guinea pigs and obtain the following weights in grams;

$$29, 31, 35, 28, 30, 27, 32, 31, 25, 26.$$

We want to determine the variance and the standard deviation. First we compute the arithmetic mean: 29.4 g. Let us now compute the ten squared deviations;

$$(29-29.4)^2, (31-29.4)^2, (35-29.4)^2, (28-29.4)^2, (30-29.4)^2, (27-29.4)^2,$$

$$(32-29.4)^2, (31-29.4)^2, (25-29.4)^2, (26-29.4)^2,$$

or

$$0.16, 2.56, 31.36, 1.96, 0.36, 5.76, 6.76, 2.56, 19.36, 11.56$$

from which
$$\sigma^2 = 8.24\,\text{g} \quad \Longrightarrow \quad \sigma = 2.87\,\text{g}$$

Now suppose we take a sample of ten guinea pigs, different from the previous one, and obtain the following weights in grams:

$$26, 34, 35, 28, 30, 25, 32, 35, 25, 26.$$

We compute the arithmetic mean: 29.6 g. We proceed, as before, to compute variance and standard deviation:

$$\sigma^2 = 15.44, \qquad \sigma = 3.93$$

while the standard deviations are:

$$12.96, \ 19.36, \ 29.16, \ 2.56, \ 0.16, \ 21.16, \ 5.76, \ 29.16, \ 21.16, \ 12.96$$

We see that although the guinea pigs of the two samples have similar weights, with almost the same arithmetic mean, the two statistical variables have very different variances. If we want to use our guinea pigs for a study, for example, on the effects of a drug, which depends on the weight of the individual, we see that the variance of the statistical variable "weight" among the individuals of the first sample is definitely lower than the variance of the statistical variable of the second sample. Hence, the first sample will lead to a more accurate experiment, as the weight of individuals is more "uniform," in the sense that it is closer to the arithmetic mean of the weights in the sample. It is important to notice that the mean is almost unchanged in both cases: this number, unfortunately, does not give us any information on how the statistical variable distributes its values within the sample.

7.6 QUARTILES AND INTERQUARTILE RANGE

The concept of quartile, associated to a statistical variable, is closely related to the concept of median. The median allows us to divide the sample into two equal parts. The notion of quartile generalizes this notion: we want to divide a given sample into four equal parts, called *quartiles*, according to the values of a given statistical variable. We could also divide the sample in an arbitrary number of parts; however, for clarity, we focus on the case of the division into four parts.

Let us look at an example to define this concept, leaving to the student the easy formulation for the generic case.

Example 7.6.1 In a high school class, we consider a sample with the following frequency table of the statistical variable recording the heights:

height 166 cm, absolute frequency 1

height 168 cm, absolute frequency 3

height 169 cm, absolute frequency 5

height 170 cm, absolute frequency 4

height 172 cm, absolute frequency 3

height 173 cm, absolute frequency 1

We want to compute the *quartiles*. We first find the median. In this case, the sample contains 17 measurements, which we sort in ascending order. As we saw in our previous section, the median is the value of the ninth measurement: $y_9 = 169$ cm. The median is the value of the *second quartile*, which we denote by writing $q_2 = 169$. We recall that in the event that the sample contains an even number of values N, the median is the arithmetic mean $(y_{N/2} + y_{N/2+1})/2$. However, the second quartile is always a value in our statistical variable. Hence, if the sample has an even number of elements N, the second quartile is *not* the median, but the value is $y_{N/2+1}$. So, for example, if we have a sample of 16 elements, the second quartile is given by the value y_9.

Now, going back to the example with 17 values, to obtain the first and third quartile, we further divide the two parts, that is, the first 8 and the last 8 values, in two parts each. We then take $y_5 = 169$ as the value for the first quartile, and $y_{14} = 172$ as the value for the third quartile.

We define *interquartile range* as the difference between q_3 and q_1: $\Delta = q_3 - q_1$. In other words, the interquartile range cuts off the 25% of the lower values and the 25% of the highest ones.

7.7 NORMAL DISTRIBUTION

We conclude our discussion with a concept pertaining to probability theory, which goes beyond this study of *discrete* statistics, that is, the study of statistical variables consisting of a finite number of values.

In fact, we want to describe the *normal distribution*, which is a continuous function, of fundamental importance and ubiquitous in mathematics, especially in statistics and probability.

Definition 7.7.1 We define *normal distribution* or *Gaussian distribution* the following function:

$$f(x) = \frac{1}{\sqrt{2\pi\sigma^2}} e^{-\frac{(x-\mu)^2}{2\sigma^2}}$$

where μ is the *expected* or *average value* and σ^2 the *variance* and they are real constants, $\sigma \neq 0$.

The graph of the normal distribution is called the *Gaussian curve*.

Let us see the graph for $\mu = 0$ and $\sigma = 1$. In this case, the Gaussian curve is called the *standard normal curve*:

$$y = \frac{1}{\sqrt{2\pi}} e^{-\frac{x^2}{2}}$$

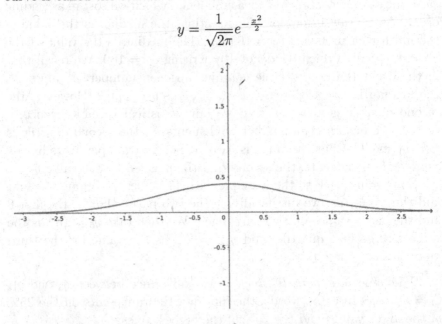

We now want to motivate why the parameters μ and σ^2 appearing in the definition of normal distribution are called average and variance, a terminology that we have already encountered for statistical variables.

Suppose we draw a histogram recording the distribution of some available data, i.e., of a statistical variable. We want to see, in a concrete example, that variance and mean value of a statistical variable are well approximated by the values σ^2 and μ, respectively, of the Gaussian curve "approximating" the statistical variable.

Example 7.7.2 Suppose we have the frequency table of the height, expressed in centimeters, of the ostriches in a farm:

Height	205	200	195	190	185	180	175	170	165	160	155
Freq.	1	0	3	2	4	4	3	3	2	1	1

We compute both the variance and the mean value, obtaining:

$$\sigma^2 = 16.58, \qquad \mu = 180$$

Now let us see on the same graph both the histogram and the Gaussian curve obtained by taking σ^2 and μ as above.

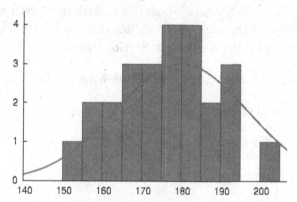

Notice that it is necessary to multiply the distribution Gaussian by the area occupied by the histogram, i.e., by the number of data multiplied by the sum of their frequencies. This is necessary because the Gaussian curve is *normalized*, that is, the area under the curve is 1.

It is then clear that in a case like this, as well as in many others, the Gaussian curve represents a good approximation of the discrete distribution of data, with the same mean and variance, allowing us to use differential calculus, which we have studied in the previous chapters.

This example represents the first step in the direction of probability theory, which uses tools of differential calculus to obtain information of statistical significance. In this text, we are unable to fully address these topics, but we would like at least to convey their capital importance in statistics.

7.8   EXERCISES WITH SOLUTIONS

7.8.1 Suppose we measure, in centimeters, the heights at the withers of ten dogs participating to a competition, obtaining the statistical variable:

$$Y = (40, 42, 38, 41, 40, 45, 46, 42, 42, 41)$$

Determine the median, the arithmetic mean and the variance.

Solution. To determine the median, we need to reorder our data, so that they are increasing:

$$38, 40, 40, 41, 41, 42, 42, 42, 45, 46$$

The median is 41.5 cm (average of the fifth and sixth value). The arithmetic mean is the sum of the values divided by 10: 41.7 cm. The variance is given by the formula in Section 7, and it is: 5.01.

7.8.2 A biology student achieves the following grades (out of 30):

Biology 1 (6 credits): 27

Mathematics (8 credits): 24

Genetics (6 credits): 30

Chemistry (6 credits): 25

Compute the weighted average, according to the credits. What happens to the weighted average if the student gets the grade of 26 in the exam of Histology (4 credits)?

Solution. We compute the weighted average:

$$(27 \times 6 + 24 \times 8 + 30 \times 6 + 25 \times 6)/26 = 26.31$$

Suppose now the students get the grade of 26 in the exam of Histology. Even without calculations, we can state that the weighted average will decrease (by a small amount), because 26 is close, but smaller than the average. Indeed, the weighted average will become:

$$(27 \times 6 + 24 \times 8 + 30 \times 6 + 25 \times 6 + 26 \times 4)/30 = 26.27$$

7.8.3 The weight in tons of some cattle is given by the following statistical variable:

$$Y = (1.4, 1.2, 1.3, 1.4, 1.4, 1.5, 1.6, 1.2, 1.2, 1.1, 0.9)$$

Compute the mean, the variance, the mode and the frequencies.

Solution. The average is 1.29, while the variance is 0.039. We have two values for the mode: 1.2 and 1.4. The absolute and relative frequencies are given in the table:

	0.9	1.1	1.2	1.3	1.4	1.5	1.6
Abs. Freq.	1	1	3	1	3	1	1
Rel. Freq.	1/11	1/11	3/11	1/11	3/11	1/11	1/11

7.9 SUGGESTED EXERCISES

7.9.1 Given the following statistical variable

$$Y = (5, 4, 2, 2, 1, 7, 4, 6, 7, 3, 3, 2, 7, 4, 2, 3, 3, 1, 5, 6, 9, 7, 5, 6, 4)$$

compute the absolute and relative frequencies of the data. Then, compute the median, the arithmetic mean, the variance and the standard deviation (mean square deviation).

7.9.2 The following statistical variable

$$Y = (7, 6, 7, 4, 5, 8, 7, 7, 8, 6)$$

corresponds to the grades for a final test of a sample of students in a high school. Compute the absolute and relative frequencies of the data. Then, compute the median, the arithmetic mean, the geometric mean, the variance and the standard deviation.

7.9.3 The following statistical variable:

$$Y = (56, 25, 34, 27, 51, 42, 39, 34, 34, 45, 36, 28)$$

corresponds to the age of the employees of a company. Compute the absolute and relative frequencies of the data. Then, compute the median, the arithmetic mean and the mode of the data.

7.9.4 A survey is carried out among 28 boys, and they were asked to indicate their favorite sport: ten indicate soccer, eight swimming, five volleyball, three tennis and two basketball. Compute the relative, absolute and percentage frequencies of the corresponding statistical variable. Draw a bar diagram of the absolute frequencies and a pie chart of the percentage frequencies.

7.9.5 The following data represent the number of insects per plant in a set of 10 plants:

$$16, 14, 21, 46, 4, 5, 12, 25, 0, 18$$

Compute the median, the arithmetic mean, the variance and the standard deviation.

7.9.6 The following data represent the cholesterol level in the blood, expressed in mg/dl, of a sample of patients at a hospital:

$$174, 138, 212, 203, 194, 245, 146, 149, 164, 209, 158.$$

Find the median, the mode, the arithmetic mean, the variance and the standard deviation.

7.9.7 The following data represent the age of patients in a hospital:

$$29, 44, 35, 36, 30, 41, 35, 45, 38, 27$$

Compute the median, the mode, the arithmetic mean, the variance and the standard deviation.

7.9.8 The following data represent the number of seeds per flower in a sample of nine flowering plants:

$$25, 38, 43, 19, 22, 33, 45, 37, 21$$

Compute the median, the mode, the arithmetic mean, the variance and the standard deviation.

7.9.9 The total energy consumption in a small town is recorded for 300 days. The data, expressed in KW, are summarized in the table:

Class	Frequency
[0, 100]	50
[100, 200]	85
[200, 400]	65
[400, 600]	55
[600, 1000]	45

The mean and the median of these data are:

A. 327.5 and 250

B. 326.9 and 250

C. 327.5 and 250.5

D. 327.5 and 251.5

Mark the correct answer.

7.9.10 For 300 days, we record the total consumption of electricity in a small municipality. The data, expressed in KW, are summarized in the following table:

Class	Frequency
[0, 100]	50
[100, 200]	85
[200, 400]	65
[400, 600]	55
[600, 1000]	45

The variance and the standard deviation of these data are:

1. 60, 871 and 247

2. 60, 870 and 246

3. 60, 870 and 245

Mark the correct answer.

7.9.11 The arithmetic mean and the median of the following distribution of the number of nurses in 23 hospitals expressed in classes:

Nurses	Hospitals
1 − 10	6
11 − 20	13
21 − 40	4

are:

A. 15.5 and 12

B. 15 and 13

C. 15.5 and 13

D. 15.5 and 11

Mark the correct answer.

7.9.12 The weights in hg of 100 newborns born were recorded in December 2020 in Ferrara by the following data:

Weight	Newborns
$27 < x \leq 30$	6
$30 < x \leq 33$	28
$33 < x \leq 36$	42
$36 < x \leq 39$	16
$39 < x \leq 42$	8

The arithmetic mean and mode of these data are:

A. 34.26 and 34.5

B. 34.12 and 34.9

C. 35.13 and 34.5

D. 33.26 and 34.5

Mark the correct answer.

7.9.13 Consider the distribution, relative to the weight of newborns, expressed in the previous exercise. The median, the variance and the standard deviation of these data are:

A. 34.5, 9.03273 and 3.00545

B. 34.5, 9.13893 and 3.00545

C. 34.9, 9.03273 and 3.00545

D. 34.5, 9.03273 and 3.10575

Mark the correct answer.

7.9.14 The following frequency table records the measurements of the height of a group of people;

Class	Abs. Freq.
166	1
168	3
169	6
170	11
171	8
172	6
173	4
174	3
175	1
178	1

The average of these data is:

A. 170.9

B. 171

C. 169.5

D. 169.8

E. None of the above

Mark the correct answer.

7.9.15 According to the table from the previous exercise, the median of these data is:

A. 171

B. 172

C. 169

D. 173

Mark the correct answer.

7.9.16 According to the table of Exercise 7.9.14, the mode of these data is

A. 170

B. 169

C. 168

D. 171

E. 178

Mark the correct answer.

7.9.17 The marks obtained by a student in written math tests were: 7, 6, 4, 6, 7. The arithmetic mean of the marks obtained is

A. 6

B. 5

C. 5.5

D. 6.5

Mark the correct answer

7.9.18 The average of the marks obtained in the first three written tests of math by a student is 5.5. In the fourth test, he gets 6.5. What is the average after the fourth test?

A. 5.75

B. 5.5

C. 6

D. There is insufficient data to compute it

Mark the correct answer.

7.9.19 The average of the marks obtained in the first three written math tests by a student is 5.5. What grade should the student get in the fourth test to obtain a mean equal to 6?

A. 7.5

B. 6

C. 6.5

D. 7

E. None of the above

Mark the correct answer.

7.9.20 The grades obtained by a student in written tests were 7, 6, 4, 6 and 7. The median of the votes obtained is

A. 6

B. 4

C. 6.5

D. 7

Mark the correct answer.

7.9.21 The median of the grades obtained in the first three written tests by a student is 6.5. On the fourth and fifth tests, he gets 5 and 7, respectively. What is the median of the five grades?

A. 6.5

B. The data are insufficient to compute it

C. 6

D. 6.25

Mark the correct answer.

7.9.22 The median of the grades obtained in the first three written tests of mathematics by a student is 6.5. In the fourth test, she takes 7.5. What can we say about the median of the four grades?

A. It definitely increases

B. Stays 6.5

C. Does not decrease

D. Becomes 7

Mark the correct answer.

Appendix A: Solutions of Some Exercises

A.1   FUNCTIONS IN APPLIED SCIENCES

1.9.1 No, yes, no, yes.

1.9.2 1) $x = 1$, no difference quotient. 2) $y = -(3/2)x + (7/2)$, $m = -3/2$.

1.9.4 1) $D = \{x \in \mathbb{R} \mid x \neq \pm 1\}$. 2) $\mathbb{R} \setminus \{0\}$. 3) $D = \{x \in \mathbb{R} \mid x > 0\}$. 4) $D = \{x \in \mathbb{R} \mid x > 0\}$. 7) $D = \{x \in \mathbb{R} \mid x \neq \frac{-3 \pm \sqrt{17}}{2}\}$. 12) $D = \{x \in \mathbb{R} \mid x \neq -3\}$. 18) $\mathbb{R}$.

1.9.7 $D(4) = (6/24) \times 480 = 120$ mg, $A = 480$ mg.

1.9.9 9.6 seconds.

1.9.12 $p = 1/2$.

1.9.13 Let $N(t)$ be the amount of ibuprofen at time t. $N(45 \text{ minutes})$ = 308.442 mg, $N(90 \text{ minutes}) = 237.841$ mg, $N(24 \text{ hours}) = 0.097$ mg.

1.9.14 4.26 years.

1.9.17 1) No. 2) Yes. $[(1.33)^2.5 \cong 2.03, (1.33)^2 \cong 1.76]$.

1.9.18 We have initially 0.000036875 g. We need 14.7 cycles to have 1 g.

1.9.19 1.068×10^{-9} g.

1.9.22 1) Linear law. 2) 3 kg.

1.9.25 Both sides are $3/\sqrt{2} \cong 2.12$.

1.9.27 The tangent is $\sqrt{24}$.

1.9.28 The tangent is $1/\sqrt{24}$.

A.2   LIMITS AND DERIVATIVES

2.10.1 2) We have $\lim_{x \to -\infty} f(x) = L$ if for all $\epsilon > 0$ exists $N > 0$ such that if $x < -N$, then $|f(x) - L| < \epsilon$.

3) We have $\lim_{x\to a^-} f(x) = -\infty$ if for all $M > 0$ exists $\delta > 0$ such that if $x_0 - \delta < x < x_0$, then $f(x) < -M$.

4) We have $\lim_{x\to +\infty} f(x) = -\infty$ if for all $M > 0$ exists $N > 0$ such that if $x > N$, then $f(x) < -M$.

2.10.2 3) For all $M > 0$, it exists $N > 0$, such that if $x < -N$, then $2x^3 < -M$. Take $N \geq (M/2)^{1/3}$, e.g., $N = (M/2)^{1/3}$.

2.10.4 1) 1, 2) 0, 3) −1, 4) −2, 5) −7, 6) 3/2.

2.10.5 1) 1/6, 3) 0, 5) −1/2, 8) −1/2, 11) −∞

2.10.6 1) 1, 2) −∞, 3) +∞.

2.10.7 1) $e^{2x}(2x^2 + 8x + 3)$, 5) $-(x^2 + 4x - 1)/(1 + x^2)^2$, 10) $[(1 + x)\sin(x) + 1]e^{-\cos(x)}$.

2.10.8 1) $y = 3x$.

2.10.9 1) $x = 0$, $P = (0, \log(3))$, 2) They do not exist.

A.3 APPLICATIONS OF THE DERIVATIVE

3.7.1 1) $\sqrt{0.98} \cong 0.99$. 3) $e^{-0.02} \cong 0.98$.

3.7.3 The height is 6.46 m. The velocity is −9.61 m/s.

3.7.8 1) Domain: $x \neq \pm\sqrt{\log(3)}$. Asymptotes: $x = \pm\sqrt{\log(3)}$ vertical, $y = 0$ horizontal. 4) Domain: $x < 0$, $x > 8$. Asymptotes: $x = 0$, $x = 8$ vertical. 15) Domain: $x \neq \pm 3$. Asymptotes: $x = \pm 3$ vertical, $y = 2$ horizontal.

3.7.9 1) Domain: $\mathbb{R}$. The function is continuous and differentiable in all the domain. 3) Domain: $\mathbb{R}$. The function is continuous and differentiable in all the domain. 6) Domain: $x < -3$, $x > 3$. The function is continuous and differentiable in all the domain.

3.7.10 1) The function is increasing for $x < 0$, $x > 2$. 2) The function is decreasing. 3) The function is increasing for $-3 < x < 0$, $x > 3$.

3.7.11 1)

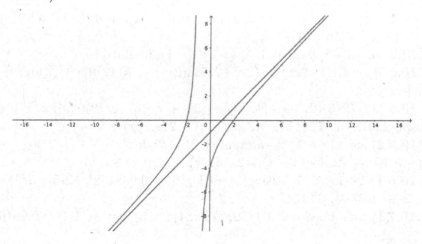

6)

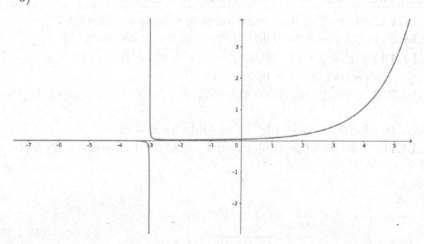

3.7.12 4 m ×6 m.

3.7.15 Objective function: $S = 2\pi rh$, r radius, h height of the cylinder $R^2 = r^2 + (h/2)^2$. We have $h = \sqrt{2}R = 30\sqrt{2}$.

3.7.21 Height is:

$$h = \frac{147 - 3R^2}{2R}$$

Volume is:

$$V(R) = -(5\pi/6)R^3 + (147\pi/2)R$$

Max occurs for: $R = \sqrt{147/5}$ e $h = \sqrt{147/5}$.

A.4 INTEGRALS

4.10.1 1) $(1/6)(2x^2+1)^{3/2}+c$. 3) $(1/2)\sin(x^2+1)+c$. 4) $(-1/3)e^{1-x^3}+c$. 5) $\log(x^3-2x+4)+c$.

4.10.2 1) $\sqrt{3}-1/3$. 2) $1/3$. 5) $(1/3)(e-1)$. 6) $1094/15$.

4.10.3 3) $(1/4)((1-2x^2)\cos(2x)+2x\sin(2x))+c$. 5) $(2/9)x^{3/2}(3\log(x)-2)+c$.

4.10.4 1) $(2/3)(2x+1)^{3/2}(3x-1)+c$. 5) $(x/2)(\sin(\log(x))+\cos(\log(x)))+c$. 7) $(1/4)x^2(2\log^2(2x)-2\log(2x)+1)+c$.

4.10.5 2) $\log(x)+c$. 4) $x-\arctan(x)$. 5) $(1/2)\log|x-4|-(1/2)\log|x-2|+c$. 8) $-(-2x\log(3-x)+2x\log(x)+3)/(9x)+c$.

4.10.6 1) $1+\log(6/5)+6\log(5)-6\log(6) \cong 0.0884$. 2) $-2(log(5/3)-6) \cong 10.9783$. 4) 0.3154.

4.10.7 1) $(e^2-1)/(2e) \cong 1.1752$. 2) 2π. 4) 2π. 5) $\log((3\sqrt{3})4) \cong 0.2616$.

A.5 FIRST-ORDER DIFFERENTIAL EQUATIONS

5.14.1 Only the last four equations are separable equations.

5.14.2 (1) No direction field. (2) Second direction field.

5.14.4 1) $y(x) = 3-e^{-x}$. 2) $y(t) = 2-e^{-t^2}$. 3) $y(t) = (1/4)(t^4+8t-5)$. 4) $y(x) = (-\cos(x)-1+\cos(1))/x$

5.14.5 b) 1) yes, 2) no, 3) no 4) yes 5) no. c) $y(x) = 2x+(x-1)\log(x-1)-3$, $x > 1$.

5.14.10 a) $C(t) = 70-10e^{-5t}$. b) Only for $t \to \infty$.

5.14.11 1) $V = 0$ (unstable) and $V = a^3/b^3$ (as. stable). 2) $V(t) = (1+ke^{-t/3})^3$.

5.14.13

1. $y(x) = cx - x^2$

2. $y(x) = -\sqrt{x^2 - log|2(c-x)|}$

5.14.14 1) $y = 0$ as. stable, $y = \pm\sqrt{3}$ unstable. 3) $x = -1$ as. stable, $x = 1$ unstable.

A.6 SECOND-ORDER DIFFERENTIAL EQUATIONS

6.11.1 1) $y(t) = c_1e^{-2t} + c_2e^{4t}$. 2) $y(t) = c_1e^{-t/2} + c_2e^t$. 3) $y(t) = c_2e^tt + c_1e^t$. 4) $y(t) = c_2\sin(\sqrt{2}t) + c_1\cos(\sqrt{2}t)$.

6.11.3 1) $y_p(t) = -e^{2t}/8$, $y(t) = c_1e^{-2t} + c_2e^{4t} - e^{2t}/8$. 2) $y_p(t) = -(6\sin(t))/5 + (2\cos(t))/5$, $y(t) = c_1e^{-t/2} + c_2e^t - (6\sin(t))/5 +$

$(2\cos(t))/5$. 3) $y_p(t) = t^2 + 4t + 6$, $y(t) = c_2 e^t t + c_1 e^t + t^2 + 4t + 6$. 4) $y_p(t) = t^2/2 - 1/2$, $y(t) = c_2 \sin(\sqrt{2}t) + c_1 \cos(\sqrt{2}t) + t^2/2 - 1/2$.

6.11.5 1) $y(x) = e^x(e^x - 2)$. 2) $y(x) = e^x(\cos(x) - \sin(x))$. 3) $y(x) = -(1/7)e^{x/4}(\sqrt{7}\sin(\sqrt{7}x/4) - 7\cos(\sqrt{7}x)/4)$. 4) $y(x) = e^x(-3x + 2e^x - 1)$. 5) $y(x) = 1/5(e^x(13e^x - 15) - 6\sin(x) + 2\cos(x))$.

6.11.6 1) $y(t) = (1/16)e^{-2x}(-3 + e^{4x}(3 + 4x))$. 2) $y(t) = -(1 + 2x)\cos(x) + 3\sin(x)$. 3) $y(t) = (1/4)(3e^{-x} + e^x(5 + 2x) - 4(2 + x^2))$.

6.11.7 $k \cong 40$, $x(t) = 0.1\cos(2\sqrt{10}t)$

6.11.9 $k = 49$, $\mu = 14$ $x(t) = -3te^{-7t}$.

A.7 ELEMENTARY STATISTICS

7.9.1 Median 4, arithmetic mean 4.32, variance 4.4576, standard deviation 2.11.

7.9.2 Median 7, arithmetic mean 6.5, variance 1.45, standard deviation 1.2.

7.9.16 A).

7.9.17 A).

7.9.18 A).

7.9.22 C).

Bibliography

[1] R. Adams, C. Essex, *Calculus: A Complete Course*, Prentice-Hall, Hoboken, NJ, 2009.

[2] W. Boyce, R. C. DiPrima, *Elementary Differential Equations*, Wiley, Hoboken, NJ, 2008.

[3] W. Rudin, *Principles of Mathematical Analysis*, McGraw-Hill, New York, 2015.

Index

absolute frequency, 201
algebraic
    function, 9
antiderivation, 109
arithmetic mean, 206
associated homogeneous
    equation, 177
asymptote, 81
    horizontal, 82
    oblique, 82
    vertical, 82
autonomous equations, 151
auxiliary equation, 179
average, 205

bar graph, 204

Cauchy initial value problem, 136
Cauchy problem, 136
Cauchy's Theorem, 170
Cauchy's theorem, 136
chain rule, 54
characteristic equation, 179
codomain, 1
composition of functions, 54
concavity, 84
continuity, 46
cosine, 16
cost function, 85
critical damping, 192
critical point, 76
cycle, 14

Damped harmonic motion, 190
definite integral, 103
derivative, 49
    cosine, 53
    exponential function, 53
    logarithic function, 53
    sine, 53
determinant, 198
differential, 72
differential equations, 135
    existence and uniqueness
        theorem, 136
    Cauchy problem, 136
    first order, 135
    general solution, 135
    in normal form, 136
    initial condition, 136
    Initial value problem, 136
    integrating factor, 143
    interval of existence, 137
    ordinary, 136
    partial differential
        equations, 136
    particular solution, 135
direction field, 138
domain, 1
    algebraic functions, 10
    exponential function, 10
    logarithmic function, 12
    rational functions, 10

Printed in the United States
by Baker & Taylor Publisher Services